IBERO-AMERICANA: 52

THE YUKPA CULTIVATION SYSTEM

A Study of Shifting Cultivation in Colombia and Venezuela

THE YUKPA CULTIVATION SYSTEM

A Study of Shifting Cultivation in Colombia and Venezuela

BY

KENNETH RUDDLE

UNIVERSITY OF CALIFORNIA PRESS
BERKELEY · LOS ANGELES · LONDON
1974

IBERO-AMERICANA: 52

Approved for publication February 16, 1973

UNIVERSITY OF CALIFORNIA PRESS
BERKELEY AND LOS ANGELES
CALIFORNIA

UNIVERSITY OF CALIFORNIA PRESS, LTD.
LONDON, ENGLAND

ISBN: 0-520-09497-2
Library of Congress Catalog Card Number: 73-78557

PRINTED IN THE UNITED STATES OF AMERICA

CONTENTS

ILLUSTRATIONS AND TABLES

Tables

Maps

Figures

Plates

PREFACE

THE RESEARCH on which this study is based could never have been done without the generous assistance of numerous individuals and organizations. The fieldwork, which was conducted among the Yukpa–Yuko of Venezuela and Colombia over a period of 17 months (January–November, 1969; June–September, 1970; and June–August, 1971), was largely underwritten by a Ford Foundation grant from the Latin American Center of the University of California, Los Angeles. Financial support was also received from the Creole Foundation of Caracas. This study was carried out in connection with the *Venezuelan Indian Project,* sponsored jointly by the Latin American Center of the University of California, Los Angeles and the *Centro Latinoamericano de Venezuela (C.L.A.V.E.).*

Many scholars in Colombia, Venezuela and the United States of America gave unstintingly of their time. In particular I am indebted to the following persons and their staffs for the preparation and identification of botanical specimens: Hugh Cutler, Howard Scott Gentry, Professor J. M. Idrobo, Tobias Lasser, Luis Mora, Enrique Pérez, Jonathan Sauer, S. G. Stephens and Lorenzo Uribe. A vouchered collection of most of the field specimens has been deposited in the Colombian National Herbarium in Bogotá.

I am deeply grateful to Dr. Juan Guevara (formerly of the *Instituto Agraria Nacional*), Mr. George Hall (Creole Foundation), and Dr. Roberto Lizarralde (*Universidad Central de Venezuela*), all of whom are residents of Caracas, for their invaluable general assistance and orientation during the early days of preparation and planning in that city. The Capuchins Romualdo de Renedo, formerly of Los Angeles del Tukuku, and Hermenegildo Pérez de Bigastro, formerly of the mission at Socorpa, kindly provided logistical support and bases for operation in the lower Tukuku Valley and the Socorpa region respectively.

David Niddrie read the manuscript and made basic suggestions regarding clarity and style. My teachers and colleagues Joseph Spencer, Jonathan Sauer and Johannes Wilbert also subjected the manuscript to close scrutiny. This monograph owes much to their advice, criticism and encouragement.

Marshall Durbin and Haydeé Seijas, linguists at Washington University, and the *Instituto Venezolano de Investigaciones Científicas,* respectively, patiently reworked my somewhat specialized glossary on the basis of their field experience among the Yukpa. Their orthographic note and revision of my glossary are reproduced here *in toto*. I am greatly indebted to them both.

I am fortunate to have had the help of a very competent assistant, Kathleen Barrows, who cheerfully handled the more pedestrian chores associated with an undertaking of this sort. I am most grateful to her. The maps and diagrams were skillfully produced by Noel Diaz. The photographs were taken by the author.

K.R.
Los Angeles, California

CHAPTER I

INTRODUCTION

THE TERM "shifting cultivation" and its many synonyms embraces numerous forms of cultivation, which, taken as a whole, have long been recognized as the most widespread and enduring land use system in the tropical world.[1] Shifting cultivation has been in use since the Neolithic era in many of the world's forested areas, having spread progressively across most of southern and eastern Asia, Europe, and humid Africa during the early stages of their settlement by agriculturalists (Spencer, 1966:4). Shifting cultivation was widespread throughout the Americas in pre-Columbian times and was used in eastern North America by the early European colonists. It seems to have been little used, if at all, in grasslands and the very arid zones.

Minimally defined, shifting cultivation is a system of continuous farming in which impermanent fields are cleared with fire and are cropped for fewer years than they are fallowed.[2] Apart from affording a minimal definition, it is of little value to present a list of those elements universally associated with this type of cultivation system, as the specific forms manifested by shifting cultivation differ considerably depending upon such variables as: the biophysical and sociocultural environments; principal crops cultivated; crop associations and successions; time ratios of cropping-fallowing; spatial distribution of cultivated fields; use of domestic animals; techniques of land use; tool inventories; and the role of ancillary activities such as hunting, fishing, collecting, animal husbandry, fixed-field agriculture, cash cropping and other occupations associated with the so-called "modern sector" of the economy.

Over most of the extra-tropical world, shifting cultivation has been replaced by advanced cropping systems which are more in balance with landscapes that have been modified and developed by human occupancy over a long period of time. It is generally agreed that the substitution of

[1] I have consistently used the terms "shifting cultivation" and "cultivator" in this monograph, rather than "shifting agriculture" and "farmer" or "agriculturalist." The term "cultivation" implies a simpler manner of working the land than does "agriculture." It is therefore more appropriate in the context of the study.

[2] It is important to note that fields are not "abandoned" as is suggested by some observers. Field sites remain as an integral part of the system with the end of cropping and the start of fallowing.

advanced agricultural systems depends upon the overall development of group culture and the evolution of more elaborate cultural landscapes.

Shifting cultivation has remained the principal agricultural system employed in the tropics, where it retains importance because of the vast territory covered and the diversity of peoples who use it. It often continues to be the standard practice where group cultures have not advanced, where the natural environment has only marginally developed into a cultural landscape, and where a group's traditional practices have been disrupted by the exploitative activities of an alien culture. Pantropically shifting cultivation is still the dominant system of the pioneer frontier; of the tribe which clings tenaciously to its traditional ways of making a living; and of such remote regions as the headwaters of the Amazon and interior highlands of New Guinea. Shifting cultivation is, however, employed by a wide variety of people under an enormous variety of conditions and circumstances.

Systems of Shifting Cultivation

In much of the early literature shifting cultivation was customarily defined in a minimal and unsystematic fashion. In gross terms it was characterized by a rotation of fields rather than of crops, in which comparatively short periods of cropping alternate with long fallows, and by fields which are cleared by relatively simple slash-and-burn techniques. Typically it was assumed that the phenomenon exhibited only minor variations throughout the tropics (Bartlett, 1956:693).

During the last two decades the degree of interest in tropical agriculture, particularly in shifting cultivation, has greatly increased. It is now realized that those early, crude definitions conceal an immense variety of subsystems, and are, therefore, of very little value. Conklin (1954), calling for a more realistic approach to the study of shifting cultivation, complained that, despite the large number of people supported by the system, it was still only inadequately understood. He stated that, "it is still often categorically condemned as primitive, wasteful or illegal, with little or no regard for ... pertinent local variables, [and there] is a definite need for ascertaining what are the real facts about shifting cultivation." In 1957 Conklin advanced a basic conceptual framework for the study of shifting cultivation, with a fuller expression of the idea of "integral" and "partial" systems. This division of systems of shifting cultivation into integral and partial types has fundamental ecological implications since it is based upon the degree of integration of the cultivation system with the other systems of a particular sociocultural matrix. Integral systems are those in which the practice of shifting cultivation forms the core of society's existence, and where the remainder of the sociocultural matrix depends upon this core. These systems form "a more traditional, year-round, community-wide, largely self-contained and ritually sanctioned way of life" (Conklin, 1957:2), whereas partial systems reflect predominantly the economic interests of the participants.

Partial systems were divided by Conklin into two basic types: "Supplementary Partial Systems" and "Incipient Partial Systems." In the former, a permanent-field cultivator, either through necessity or as a tenant, devotes part of his efforts to shifting cultivation. The latter comprises homesteaders, squatters, or settlers in an organized colony who have moved into an area of shifting cultivation from a crowded zone where permanent-field agriculture is the mode. Operators of partial systems often have little prior knowledge of the type of cultivation which they adopt in the new location. Often they are forced from necessity to adopt alien systems, which frequently become permanently supplementary. Although the operators are at least "knowledgeable as to the minimal practices and operations of the systems employed" (Spencer, 1966:2), they are more likely to cause ecological damage than are the operators of an integral system. It may be inferred that cultivators following incipient partial systems, whose knowledge of the correct practices is virtually nil, achieve a lower productivity than cultivators employing a supplementary system, and at the cost of greater ecological damage.

A major contribution to the study of systems of shifting cultivation was made by Spencer (1966) in a survey of the phenomenon in Southeastern Asia. Besides examining previous definitions and analytical criteria, Spencer provided a wealth of information extending earlier concepts. To Conklin's two types of partial system he added a third, "opportunistic," (ibid.:80), "in which individuals of many callings and primary occupational patterns temporarily or regularly carry on a variant form or version of shifting cultivation for any of several different reasons whenever the opportunity presents itself." He commented further (ibid.:81) that the practitioners of opportunistic systems (by definition of the types of individuals included) are "far more out of context (compared with integral system cultivators) and . . . form a random series of anomalous cases rather than a composite series." Hence the knowledge of conservational practices of shifting cultivation, the technology employed, the impact on the landscape, and the productivity vary enormously from type to type.

Spencer considered all classes of shifting cultivators, regardless of cultural level, and proposed the existence of seven types of partial systems (ibid.: 165)[3]: Temporary chance patterns continued without evolutionary elements; frontier migrant settler, evolutionary toward permanent-field farming; frontier zone, declining practice of former shifting cultivator; permanently complementary (subsistence complement to permanent-field farming); permanently complementary (commercial complement to permanent-field farming); urban fringe commercial (market production); and, finally, sedentary farmer with private land control (shifting cultivation permanently complementary inside farm holding).

Expansion of the Partial System Typology

No student of shifting cultivation in attempting to establish a typology of partial systems has dealt adequately with those systems which have developed from the break-down of an earlier integral system of shifting cultivation, and from the evolution of a formerly integral autosubsistence system in which shifting cultivation was no more important than a number of other sources of food. Conklin's category of partial systems needs to be expanded to accommodate the mixed economy of the incipient tropical forest horticulturalist (Leeds, 1961:5), for whom the total food quest constitutes the cultural focus of the society. Although the autosubsistence system as a whole is obviously "integral," the system of shifting cultivation which forms only a segment of the whole is a "partial" system. Various changes in the biophysical and cultural environments may be postulated by which a society of incipient tropical forest horticulturalists could evolve to the level of advanced tropical forest horticulturalists for whom hunting, gathering, and fishing would be relegated to a subsidiary role, and where the society would be organized around the activities of shifting cultivation. In such a way a hypothetical change may be envisaged from a partial to an integral system of shifting cultivation.

Conversely, one can hypothesize the evolution of the mixed economy of a group of incipient horticulturalists into a number of partial systems of shifting cultivation, in which hunting, gathering, and fishing would be relegated to the role of ancillary activities, and where shifting cultivation

[3] Although Spencer was essentially concerned with partial systems in relation to permanent-field agriculture, he did not define partial systems by this relationship.

would assume the major supportative function. In this case the ancillary activities could retain their symbolic importance, and may be emotionally favored over cultivation. Although the group would be sustained by the products of shifting cultivation, the system would not form the focus of their sociocultural organization. The social organization would be a carry-over from an earlier era, when the autosubsistence system focused upon those activities which, it is hypothesized, would be relegated to ancillary roles. This last example probably sketches the evolution of the present-day autosubsistence of the Yukpa, who operate partial systems of shifting cultivation.[4]

Successes and Problems of Shifting Cultivation

Scientists schooled in the Western tradition of resource use are almost unanimous in their a priori condemnation of shifting cultivation, which they see as a primitive, wasteful, destructive, and unproductive system of land use. Occasionally, however, it has been pleaded that under traditional conditions shifting cultivation is not axiomatically ruinous and that it provides an ecologically sound way of using land in many tropical environments which would, perhaps, suffer more damage under some ill-planned form of sedentary agriculture.

The great triumph of shifting cultivation has been its capacity to support large numbers of people through many centuries, not by an extraordinary and laborious reworking of the natural landscape, but rather through a more subtle and ingenious imitation of the natural order by a miniature tropical forest of cultivates. In so doing, shifting cultivation transforms natural forests into "harvestable" forests (Kampto Utomo, 1957:129, cited in Geertz, 1966:25).

Geertz (1966:16–25) has admirably summarized those basic positive attributes of the pantropical success of shifting cultivation. Of prime importance is the fact that the cultivated plot, with each of its great variety of intercropped species represented by a small number of individual plants, simulates the tropical forest in its degree of generalization or high diversity index. Moreover, under ideal conditions, when it is genuinely adaptive, shifting cultivation is integrated into and maintains the structure of the natural ecosystem into which it is projected. Indeed, there is a remarkable systemic congruity between the cultivated plot and the stable climax forest.

[4] The term "Yukpa" is applied to all of the subtribes throughout this study. It is synonymous with the less accurate term "Yupa" which is more widely used in the literature.

Whereas the majority of systems devised for the utilization of natural resources function by changing generalized biotic communities into more specialized ones, integral systems of shifting cultivation enable man to exploit his habitat by more or less maintaining the gross pattern of natural community while changing selected items of its content.

Secondly, both the tropical forest ecosystem and that created by integral systems of shifting cultivation seek to minimize the escape of energy and to circumvent the problems created by impoverished soils by largely feeding on the products of their own decay. Hence, in both systems, there is a high ratio between the quantity of nutrients in the biotic community to that in the substratum. The natural decay and regeneration systems of the tropical forest is replicated by shifting cultivation (burning, cultivation, decaying cultivates [after removal of edible portion of plants], secondary succession of natural vegetation) and the activity in both takes place principally in the biotic community. In the case of shifting cultivation, however, man greatly accelerates natural decay by the burning of the slashed and felled fields and a large amount of energy escapes across the boundaries of the system because the energy transfer affected by "accelerated decay" (i.e., slashing, felling and burning) is less efficient than that which takes place under natural conditions of decay and regeneration. After two or three years of cultivation this energy loss results in the loss of fertility in the fields.

Finally, in "architectural" terms both the tropical forest and the cultivated fields are "closed-over," multi-storied structures which serve to protect the soil against the damaging impact of direct rainfall and insolation. Thus the severity of potential soil erosion is lessened. Both systems control the growth of ground cover which in shifting cultivation lessens the arduous task of weeding. The apparently haphazard interplanting of species with different growth habits and root systems of the polycultural field of integral systems insures an effective vertical and lateral exploitation of available light, warmth, moisture, and nutrients while at the same time protecting the soil.[5]

Given those ecological problems which, as many "modernizers" have discovered to their chagrin, attend the human exploitation of tropical environments, and the technological resources of the autochthonous popu-

[5] See Harris (1971:475–495) for comparative "ecological fitness" of "polycultural" fields under integral systems and "monocultural" fields under partial systems in the Upper Orinoco region.

lations of most such regions throughout the bulk of their history, integral systems of shifting cultivation represent remarkably perceptive and sophisticated ways of using what, to most westerners, are difficult regions in which to sustain societies. The historical success of systems of shifting cultivation as rational systems of resource use is unquestionable, for until the last century and, indeed, until the last two or three decades in many cases, they have sustained the bulk of the human population in most tropical regions.

While most systems of shifting cultivation may in the past have represented sound adjustments to biophysical conditions, however, it is becoming increasingly evident that, owing to the explosive growth of world population, such systems are no longer rational. Under less than ideal conditions (generally those impinging upon partial systems but increasingly upon integral systems) shifting cultivation may become maladaptive. The equilibrium of the cultivated field is far more delicate than that of the tropical forest which it imitates, and as such it is highly susceptible to breakdown leading to processes of severe ecological deterioration. Maladaptive systems have led to those problems which have provoked the negative attitudes usually applied indiscriminately to all systems of shifting cultivation.

The rapid growth of population in tropical Latin America during the last few decades has adversely affected systems of shifting cultivation in a variety of ways, which, for convenience of presentation, may be grouped in the following manner: Attempted intensification of land use by the reduction of crop-fallow time ratios; intensified "land hunger" leading to an increased number of cultivators using partial systems; the breakdown of integral tribal systems owing largely to the alienation of their ancestral lands by non-tribal populations and by the desire of some tribe members to participate in the regional market economy; and, finally, the attempts by governmental and other official institutions to modernize shifting cultivation by substituting more intensive systems of land use frequently through the wholesale introduction of agricultural technologies hitherto untested in specific sociocultural and biotropical milieux.

The most visible damaging effects of shifting cultivation have resulted from the extension of the period of cropping and the reduction of the fallow time as a consequence of greatly increased local population pressures. When traditional cultivation systems are called upon to support a continuously increasing number of families without a parallel increase in the land areas cultivated there is no alternative but to reduce the crop-fallow time ratio. Watters' (1971:30) work in Venezuela shows that crop fallow time ratios of 2:4 or 3:5 or less almost certainly indicate a progressive deterioration in soil conditions. Ever increasing and incessant demands placed upon a soil with decreasing fertility levels often results in virtual sterility of the soil and irreversible damage to vegetation.[6] These conditions lead to greatly accelerated gully and sheet erosion, particularly where excessively steep slopes are cultivated, and to the loss of irreplaceable topsoil. This in turn leads to the disruption of the hydrologic system and results in the serious flooding, silting, and eventual desiccation of the lower portion of river basins.

Partial systems of shifting cultivation employed on the pioneer frontier generally have malignant results similar to those described above. In most instances pioneer cultivators using shifting systems have been forced out of long-settled areas by "land hunger." During the course of their migration toward some undetermined goal they become transient settlers in a series of localities which are both new and, in ecological terms at least, generally unknown to them. The net result is that traditional systems which were environmentally sound in the source region of the migrants rapidly become maladaptive in the zones of in-migration,[7] a phenomena which is exacerbated by the settlers' lack of emotional attachment to the new lands being worked.

Land hunger among the non-tribal population and the thrust of pioneer cultivators deeper into the interiors of most Latin American nations has increased pressure on the Indians' ancestral lands, and is an ever more important factor in the breakdown of formerly integral tribal systems of cultivation. There is also an increasingly visible trend among some tropical forest Amerind groups to discard their integral systems and adopt partial systems in order to participate more effectively in the regional market economy. This is clearly exemplified in many tribal areas of Venezuela where the cultivation of polycultural fields has become of secondary importance to the monocultural production of maize, manioc, and plantains

[6] This is especially true when fields are either so large or so concentrated that soil recovery cannot be assisted by seeding from the surrounding forest. Further, an impoverished vegetation hastens the degeneration of the soil by permitting the unimpeded impact of sun and rain directly upon the surface of the exposed soil.

[7] This is very well exemplified in the so-called lowland "Zones of Influence" to the north and south of the Venezuelan Andes which have been settled by *andinos* migrating downslope.

for market. Monocultural fields, as previously discussed, represent a less stable system for using the tropical forest environment.

The biggest single threat to the long term survival of shifting cultivation is the criticism that it is an inefficient and wasteful extensive system of resource use, which is incapable of satisfying the demands of a rapidly increasing and impoverished population for foodstuffs and industrial raw materials. There is really very little that one can say to counter the argument that even ecologically sound systems of shifting cultivation tie up too much land in a low level of production except to point out in a theoretical way that, properly managed, integral systems of shifting cultivation, suitably modified by the introduction of selected elements of temperate zone technology and improved crop varieties, may perhaps be adapted to become environmentally sound production systems capable of supporting large, contemporary urban populations. For despite the developers' condemnation of shifting cultivation, the system's ecological effects may be no more malignant than those of the more "advanced" alien agricultural technologies when they are indiscriminately applied.

Moreover, the developers' minimal knowledge of the rational use and conservation of renewable resources in the tropics is exceeded only fractionally by their almost total ignorance of enduring autochthonous and peasant systems of resource use. Tribal autosubsistence systems in particular are disappearing with increasing rapidity as the standardized monotony of Westernization disrupts and often destroys traditional cultivation practices, some of which may have much to teach Western man about the management of fragile tropical ecosystems. The study of such systems is urgently needed.

Little time remains in which to conduct that research. All of the tropical regions of South America are threatened with the specter of development projects, many of which are basically ill-adapted to the tropical milieux. The gravity of the possible ecological disaster is exemplified by events in the vast Amazonian territory, where conservative estimates suggest that, given the continuance of present and projected trends, all of the immense tracts of rain forest will have been converted to grassland, scrub savanna, and second growth forest within a century. Richards (1970:199) is more pessimistic. He predicts the disappearance of most of the undisturbed Amazon rain forest within the next 28 years. Smaller and less remote regions of the continent will be changed even more rapidly. A more thorough knowledge of viable alternative systems of tropical lowland resource use should help prevent the realization of these alarmist predictions of an ecological disaster of monstrous proportions. It will certainly better equip us to confront the challenges of the less productive ecosystems that will replace the rain forest.

Shifting Cultivation in South America

Shifting cultivation, known variously in South America as *agricultura nomada, chacra, conuco, roça,* and *roza y tumba,* is the principal system of agricultural land use in almost all of the humid tropical areas of the continent. Paradoxically, despite the almost universal condemnation of this system, its use is becoming more widespread throughout South America as pioneer frontiers are pushed deeper into hitherto remote regions. Indeed, it is becoming increasingly apparent that the growth of agricultural production of tropical Latin American countries is only in small measure the result of the application of improved agricultural technology. The increased output is mostly the result of a considerable extension of the area worked by various forms of shifting cultivation.

Unfortunately, compared with other tropical regions, tropical South America suffers from a virtual absence of research into the nature of its contemporary subsistence systems (Conklin, 1963). So far as farming systems alone are concerned, this lacuna is particularly visible in the realm of shifting cultivation. The noteworthy contributions of Carneiro (1960, 1961), Leeds (1961), and of Watters (1971), among others, do not compare in scope and specialized content with those of Allan (1967) and de Schlippe (1955) in Africa; Barrau (1958 and 1961) in Oceania; and Conklin (1957), Freeman (1955), Izikowitz (1951) and Spencer (1966) in Southeastern Asia.

The need for this type of study among South American societies is made more urgent by the rapid growth of population in the area, and by the dependence of a large percentage of these people upon shifting cultivation. Of the 200 million people (estimated, F.A.O.:1957)[8] who derive their subsistence wholly or partly from shifting cultivation, 65 million, inhabiting almost 2.25 million hectares, live in tropical America (Nye and Greenland, 1965:7).

All of the South American countries with humid tropical regions have populations of Amerind shifting cultivators. Some tribes continue to operate their traditional integral systems whereas others, heavily influenced

[8] That is, approximately 7 percent of the estimated total world population in 1957.

by the general rural ways of the respective countries, now practice partial systems. Throughout the continent the operators of partial systems constitute the most widespread type of shifting cultivator. Three broad categories of partial systems are found throughout most of the non-tribal areas.

The first is that of the colonist advancing with the pioneer frontier. Such systems are operated throughout South America by groups of colonists such as those who settle spontaneously along newly completed highways through forested zones. These systems are often particularly destructive since this type of cultivator, who is essentially a product of local and national economic and social circumstances, is generally extremely mobile and is always seeking to profit from the development of new infrastructures. The second category consists of shifting cultivators linked by some form of agreement with large-scale landowners. This category is elaborated in the following section which deals with shifting cultivation in Venezuela. The third category of partial systems is that used by the independent cultivator in the longer settled areas. In South America this category is very heterogeneous and generally comprises the use of shifting cultivation as a complementary activity to sedentary field agriculture. Other cultivators nently in the urban areas and derive their subsistence from food raised by during the slack season of the cultivation cycle, whereas others live permanently in the urban areas and derive their subsistence from food raised by shifting cultivation on plots only a short distance from the metropolitan boundaries.

Generally partial systems of shifting cultivation are destructive of soil, forest and hydrological resources, particularly since the Latin American population explosion has caused greatly increased local pressures on available land, which, in turn, have led to a disastrous shortening of the fallow period. Throughout South America, however, the independent shifting cultivator in the longer settled areas is beginning to participate in the market economy and is gradually adopting sedentarized agriculture systems involving the cultivation of such perennial crops as coffee, bananas, cocoa, and pig and cattle raising.

Shifting Cultivation in Venezuela

From my own field observations in Brazil, Colombia, Ecuador, and Venezuela, it is apparent that in terms of the variety of systems of shifting cultivation present, Venezuela may serve as a microcosm for the whole of South America. In Venezuela, as in most other countries of contemporary tropical Latin America, the shifting cultivator[9] represents the peasant or traditional part of the dual economy that comprises the agricultural sector. Perhaps as many as 60 percent of the total number of farm units in Venezuela are worked by some form of shifting cultivation.

Broadly speaking, the systems used in Venezuela are comparable with those in use throughout South and Middle America. From personal observations and a perusal of the literature (especially useful were Chaves [1962 and 1963] and Petriceks [n.d.]), I have prepared the following general list of systems found in Venezuela.

Integral Systems. In Venezuela integral systems are now used only in those peripheral areas of the country occupied by Amerind cultivators. Not all Amerind cultivation systems are, of course, integral. Watters' (1971:55) inclusion in this category of those *criollo*[10] cultivators whose entire agricultural production is derived from shifting fields seems to me to be completely unjustifiable, as such systems manifest few of the diagnostic characteristics that distinguish integral systems, and, indeed, in Venezuela and much of Latin America are only transient forms typical of the pioneer frontier.

Partial Systems. Three broad classes of *conuco* operated under partial systems are found in Venezuela: The first is found in the Caribbean Coastal lowlands and the region known as Barlovento, where shifting cultivation is practiced alongside (and is often subordinate to) the cultivation of cacao. In these areas cacao and shade trees are planted in lieu of fallowing after the cessation of cropping. There also exists in this region a form of shifting cultivation devoted primarily to the production of Venezuelan staple foods for market. Of secondary importance is cultivation for family subsistence and pig raising.

The second class of shifting cultivation is found in the *Llanos* region. It takes place in gallery forests and exists in intimate association with the large, and sometimes highly technified farms known as *hatos*. In this region pig raising is always an important complementary activity. Three subclasses are distinguished: Those on cattle ranches where the *conuco* is cultivated by employees of the *hato* with the object of supplying both the

[9] In Venezuela shifting cultivators are known as *conuqueros;* the field(s) is a *conuco.*

[10] The term "*criollo*" is used in Venezuela to denote a native-born citizen of the country. Formerly it had biological, social and historico-cultural implications, but nowadays it seems to be applied indiscriminately, and with a sense of pride, to distinguish the native-born Venezuelan from the many naturalized immigrants. The term is not commonly used in Colombia.

owner's family and employees with subsistence crops; the system of *peguajal* or *acuaro,* in which *colonos* or *vecinos* use the land of the *hato* to make a *conuco* and then repay the landlord with seasonal labor; and finally, those where, unlike the *colonos* or *vecinos,* some independent shifting cultivators pay a cash rent in order to make a *conuco* on *hato* land. Under this system a cultivator seeks returns from a cash crop (such as cotton around Unare, and sugarcane with tobacco in Apure) in addition to fulfilling his subsistence needs.

The third class of *conuco* is found in the Andes, Coast Range, and Southern Zulia, where shifting fields are worked alongside farms operated by a rudimentary form of sedentary agriculture. They also abut against purely market-oriented production systems. In the Andean sections of the states of Barinas and Portuguesa, the basins of Uribate and Caparo, on the northeastern slope of the Sierra de la Culata, and in the Serranía de Bergantín shifting cultivation operates alongside the commercial production of coffee. Throughout the mountains of the interior of north central Venezuela and in the Serranía de Nirgua commercial cattle raising and coffee production on sedentary fields are closely intermingled with the plots of shifting cultivators. In the Andes-Llanos transition zone *conucos* on the spurs of the Andes interdigitate with *hatos* producing commercial crops located on the gently undulating floors of the valleys.

In the southern portion of the state of Zulia the system known locally as *"la fundación"* is particularly well developed on *haciendas* devoted to cattle production. It is also common throughout the serranias of interior North-Central Venezuela. Under this system a shifting cultivator contracts with an *hacendado* to make a *conuco* of a certain area, rent free, for a specific number of years. At the end of the period he is obliged to clean the plot thoroughly and sow pasture grasses before starting clearance at a new site.

In the Andes shifting cultivation is confined to the humid subtropical slopes below an altitude of 1600 m., where it is found in rugged country on slopes of at least 50 percent. These are areas with a relatively light population density of less than 20 persons per square km. Traverses by Watters (1971:68) in a number of major Andean valleys show that maize, manioc, and black beans, often associated with coffee, are cultivated to an altitude of approximately 1000 m. Between 1000 m. and 1600 m., plots are given over more exclusively to the cultivation of coffee, with sugarcane in the fertile valley bottoms. At about 1600 m. shifting cultivation is gradually replaced by sedentary agricultural systems under which truck crops occupy the best soils and wheat, oats, barley, and potatoes are grown on the intermediate soils. Livestock is grazed on the imperfectly drained, intrazonal Alpine Meadow soils.

There are in Venezuela, as throughout tropical South America, many distinct types of cultivation grouped together under the generalized rubric of "shifting cultivation." The failure of many government and international organizations to distinguish the various systems in Venezuela has, unfortunately, resulted in considerable confusion, waste, and frustration in efforts to "develop" the *conuquero*.

Shifting Cultivation among the Yukpa

The partial system of shifting cultivation practiced by the Yukpa has undergone continual change since early post-Conquest times, when the Indians were forced from their traditional lowland habitat into the alien highland world which forms the core area of their present territory. At the present time changes are taking place with increasing speed as the Yukpa adopt traits characteristic of the *campesinos* of Colombia and Venezuela.

Historically the Yukpa appear to have manifested many of the characteristics of incipient tropical forest horticulturalists, and to have operated an autosubsistence system in which a partial system of shifting cultivation was of no greater importance than hunting, fishing, and gathering activities. Despite technological adjustments to the cultivation system, it became maladaptive on the steeply sloping terrain of the new habitat and the physical environment has suffered and continues to undergo widespread damage and anthropogenically induced change. Although shifting cultivation gained in importance as other autosubsistence activities declined to an ancillary level owing to an impoverished resource base, Yukpa society apparently made few concomitant changes, and the cultivation system never assumed "integral" characteristics. Now the Yukpa are in the process of discarding their "relict" ancillary activities (hunting, gathering, and fishing) and adopting such "modern" complements to cultivation as cash cropping, wage labor, and cattle ranching.

Among the contemporary Yukpa subtribes the importance of shifting cultivation varies considerably, although its products provide the mainstay for the majority of the tribesmen. For the Japreria hunting and gathering continue to play the major role in subsistence and the cultivated field remains of secondary importance. Among the Iroka, Rionegrinos, Wasama, and Parirí the reverse is true. A more complicated range of activities is found among the Irapa. Many of the more isolated families in the headwaters of the Tukuku remain avid entomophagists, gatherers, and hunters, whereas some families are entirely dependent upon the products of cultivation and undertake virtually no ancillary labor. In the lower valley some families have ceased to be shifting cultivators and now earn their living by the production of cash crops and cattle ranching. Indeed, this change has now gone so far that some of the more acculturated Yukpa utilize the *peguajal* or *acuaro* and *"la fundación"* systems (described above as typical of the large *haciendas* of western and southern Zulia) in which they employ the labor of more traditional members of their own extended families who have recently migrated from the interior of the highlands.

Investigation among the Yupka

Although the Yukpa have been the subject of more than 100 publications only a very small percentage of these are scientific in nature. Unfortunately many of them are narratives of the "daring explorer" genre, replete with more fiction than fact. Many, too, are short articles which have been published in the monthly journal of the Capuchins, *Venezuela Misionera,* which are useful in providing numerous, though somewhat biased, insights into the Perijá region and its inhabitants.

For the purposes of this rather specialized study only very few of the earlier publications were found to be of value. One of the first writers of significance to enter Yukpa country was Jahn, who, from 1909 to 1922, made several short trips into the western Maracaibo lowlands. In 1919 de Booy lived for six weeks among the Macoa (Macoita), and explored the headwaters of the Apón. During 1914, and 1918, and again in 1920, Bolinder visited several subtribes of the Yuko. In 1937 he crossed the mountains from Colombia to Venezuela. In 1945 Reichel-Dolmatoff led a brief expedition to the Iroka, and shortly thereafter the *Sociedad de Ciencias Naturales La Salle* (*S.C.N.L.S.*) began a series of studies is the Negro Valley. Short trips were made by several *S.C.N.L.S.* high school groups during the Christmas-New Year periods of 1947–48, 1949–50, and 1950–51. In 1959, Wilbert spent several weeks among the Parirí and Wasama subtribes.

Methodology

Ideally, studies of the complex range of shifting cultivation systems should be based on painstaking microgeographic and interdisciplinary studies of every type of system, and should utilize the latest techniques of the various disciplines which impinge upon the subject. Such studies should be developed piecemeal (but with a common structure) from studies in depth of the holdings of single cultivators, because in partial systems communities may well be fairly heterogeneous. Only in this way can the great diversity of partial systems and their various and often complex technologies be truly appreciated. Logically, too, such basic identification of resource use systems should precede thorough ecological studies. With this in mind I undertook this study of the Yukpa autosubsistence system.

The Maracá subtribe was selected in order to provide a baseline against which other subtribes could be gauged, because it has retained the bulk of its traditional culture. The baseline was later extended to encompass the conservative Irapa groups living in the upper Tukuku Valley. Examples from the heavily acculturated groups were selected among the Irapa living in the lower Tukuku Valley and in the Valley of the Motilones. Additional data were gathered from the acculturated Rionegrinos inhabiting the settlements at Ajayapayna and Manastará. Because of the differential impact of acculturation, the reader is cautioned that additional studies are required before this material can be indiscriminately applied to other subtribes.

One of the universal problems which has beset students of autochthonous societies is the location of the study areas in regions which are often difficult of access. Because these areas are generally low on the nation's list of priorities, they seldom have cartographic or aerial photographic coverage. For the Yukpa country preliminary survey maps made by various oil companies are available for the Venezuelan flank of the mountains, as are preliminary topographic maps, made by the *Instituto Geográfico "Agustín Codazzi,"* for the Colombian section. Unfortunately, these maps, because of their preliminary nature, are often inaccurate, and their scale is too small to be of use in the type of study being made. Furthermore, because the study area is located across an international border, it was difficult to obtain the release of aerial photographs.

In the absence of large-scale maps, and of course, without the aid of cadastral surveys, first priority was given to surveying and map making. The whole of the Maracá community was mapped, as was the land of selected cultivators in the lower Tukuku and Negro Valleys. Mapping was done with a Brunton compass and a metric tape. All fields were measured by making a transect around the perimeter, and then by plotting the results on graph paper. The plotting which was done in the evenings immediately after the survey had been made, and during rainy periods, was crucial to the study. Not only does it permit a close check on the accuracy of the survey, but it also provides a rapid method of measuring the size of fields, which forms a basic part of economic studies among cultivators. Furthermore, plotting on graph paper permits a more accurate random sample of field-crop plants, which is also basic in making an economic analysis of the cultivation system.

No hostility was encountered during the course of this basic survey work, nor indeed at any time during the course of the fieldwork. Besides the aid of a number of young and able Yukpa who willingly acted as paid assistants, the general curiosity of the people soon turned into helpfulness, and they insisted on showing me every aspect of their valleys.

Although more than three months were spent on this wearing and often frustrating survey work, its value proved to be much more than the mere production of a land use and cadastral map. While making detailed measurements around the perimeter of every crop association, no matter how small the area involved, I soon obtained an intimacy with the minutest details of land use that could not possibly be gained in any other way. This familiarity proved to be vitally important when it came to cross-checking the data of informants.

An understanding of the various cultivation and land use systems of the Yukpa was obtained by using a number of complementary techniques. While accompanying cultivators during the various phases of the cultivation cycle, I was able to observe, participate in, time, and photograph the varied activities. After a basic knowledge of the cultivation systems and land use was obtained, and after particularly close frendships had been established with certain trusted informants, I began the process of formal, extended interviewing. The data obtained were cross-checked with other informants, and were verified by my own observations.

Most studies of tribal autosubsistence systems tend to ignore the vitally important data on labor requirements and productivity. This is not surprising, because they are not generally easy to obtain. The Yukpa lack the basic statistical concepts which we take for granted. The Yukpa numerical system does not go beyond 40 in exact terms.[11] Similarly, concepts of linear scale range from *tíuš* (very small) to *misíni* (big); area ranges from *tikémi* (narrow) to *kuwáčrič* (very wide); and weight from *ikátnora* ("hardly any weight") to *sarnúwiraysa* ("too heavy to lift"). It is therefore of very little use to ask informants to give, or to record, quantitative data.[12] All measurements have to be obtained by dint of repetitive, tedious labor on the part of the fieldworker. The activities of the various cultivation phases must be timed, the areas covered during the time period measured, and harvested products weighed.

My method of measuring crop production varied according to the crop. Stems of plantains and bananas were weighed using a spring type baggage scale. Certain cultivators were provided with standard-sized sacks holding about 45.5 kilos of grain in which to store maize and beans. This permitted the total production of a field to be more readily assessed. The weight of individual manioc and yautia plants was taken, and additional data were obtained by weighing the total root crop production of selected sample quadrats within the fields.

One investigator working alone on this type of study cannot hope, within the space of 18 months, to produce a full range of quantitative data which is statistically valid. I was able to be in only one place at a given time, therefore my attention was concentrated intensely on 17 cultivators, and the bulk of my quantitative data was derived from their fields. The land of 39 other cultivators was sampled when time permitted.

[11] One through 40 can be counted by using one's own fingers and toes, and those of the person to whom one is counting. For example, *kúmaš* means "one" and also "one finger" and *píkat óšipok* (forty) means literally "I've run out of your toes on which to count." An amount greater than 40 is called *tomónči*, "much or many."

[12] The literate Yukpa from El Tukuku were, for various reasons, found to be unsuitable.

CHAPTER 2

THE SOCIOCULTURAL ENVIRONMENT

IT IS NOT my purpose here to discuss at length the gradual evolution of ideas concerning the ethnographic composition of the Colombian-Venezuelan border region.[1] In view of the extremely confused nomenclature, however, and the variety of tribal and subtribal names which have been applied to the autochthonous population of the region, some introductory comments are merited.

TRIBAL IDENTIFICATION[2]

From the earliest days of travel and colonization in the area west of Lake Maracaibo the term "Motilones"[3] was applied indiscriminately to all of the Amerind inhabitants who wore a cropped hairstyle. The term was apparently first popularized in the seventeenth century by Simón (1882–1892) and its use has continued, with qualifications, until the present day.

During the early period of missionary activity (1779–1792), it became clear that a distinction should be made between "Chaques" and "true Motilones." In the Province of Maracaibo the Capuchins distinguished the various subtribes of the "Chaques" (Yukpa) living in the Sierra de Perijá. The northernmost were the Sabriles, who occupied the Socuy Valley. Southwards lived the Coyamos and the Macoaes, in the Palmar and Apón Valleys respectively (Alcácer, 1962:15), whereas further south dwelt the Aratomos.[4] The Chaques inhabited the area south of the River Apón.[5]

West of the mountains the "Chaques" were known as the Tupes.[6] The Coyamos extended into the Province of Santa Marta where they were known also as Tocaimos. Southwards, in the neighborhood of Becerril lived the Acanayutos and Pampanillas.[7] The priests knew these to be subgroups of the tribe now called the Yuko. The subtribes, on both sides of the mountains, were recognized by the Capuchins as belonging to one unit, namely the present-day Yukpa. All other "Motilones" were excluded from

[1] See Wilbert (1961) and Ruddle (1971) for an elaboration.

[2] In the last part of this section I have made extensive use of Wilbert's paper: *Identificación étno-lingüística de las tribus indígenas del occidente de Venezuela* (1961).

[3] The term "motilones" means "the short-hairs." It derives from the Spanish verb *motilar*, "to cut the hair short."

[4] Archivo Nacional de Colombia (A.N.C.), Botgotá, *Conventós*, t.32 f.253.

[5] A.N.C., *Conventós*, t.32 f.255.

[6] Fray Mauro de Cintruénigo, Archivo General de las Indias, Seville, *Santafé*, 547.

[7] A.N.C., *Historia Eclesiástica*, t.15 f.260.

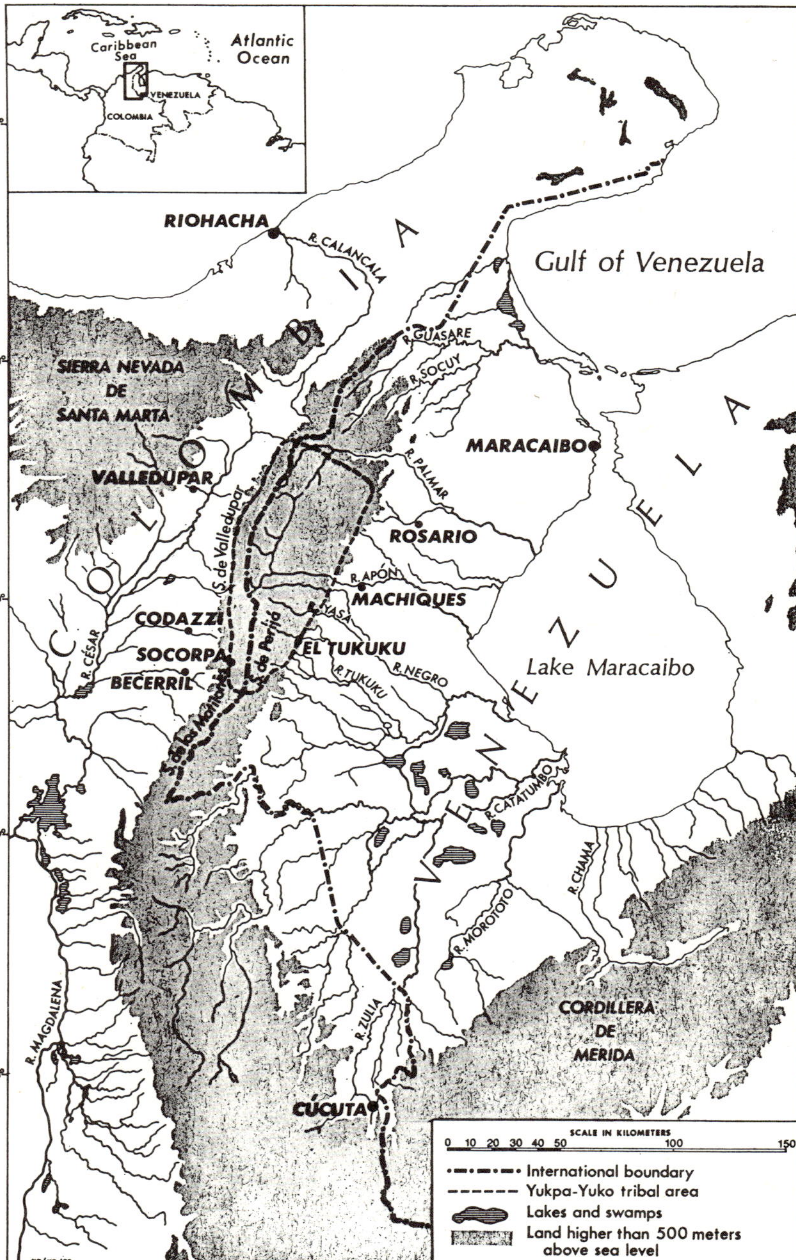

Map 1. Location of the Yukpa.

this tribe and were regarded as "true Motilones," whom we now know as the Barí.

This knowledge of the tribes was lost when the Capuchins withdrew during the War of Independence. The settlers who followed later distinguished two fundamental types of "Motilón" based upon the degree of hostility which they manifested towards the newcomers; the "Tame Motilones," and the "Wild Motilones" who lived further south.

A major advance in our knowledge of the area was made by Ernst (1887) when he recognized that the Motilón language belonged to the Cariban linguistic stock, of which the various groups spoke dialects. Jahn (1927) attempted to classify the Motilones more precisely. By substituting distributional for behavioral criteria he distinguished the Chakés (hitherto "Tame Motilones"), who lived between 10° 30′ N and 9° 50′ N,[8] from the Mapés (hitherto "Wild Motilones") who lived between 9° 50′ N and 9° N. Both groups, he said, were composed of various subgroups who spoke dialects of Motilonan, a language of Cariban affiliation. Jahn's classification was followed by Mason (1950). Reichel-Dolmatoff (1945:19) applied the term Kunaguasaya, "people of the water," to the Mapés. These he later (1960:184) tentatively divided into two distinct groups, whom he designated "Dobokubí" in the west, and "Mapé," in the east.

The ethnology of the area was further clarified by Rivet and Armellada (1950), who analyzed a rediscovered vocabulary collected by Fray Cartarroja in 1738. The analysis showed that the Mapés ("Wild Motilones"), whom they referred to as the Dobokubí, spoke a language of Chibchan stock. Thus, the tenuous behavioral and distributional criteria hitherto used to distinguish between the two Motilón groups were discarded in favor of more valid linguistic evidence. This led Wilbert (Gines and Wilbert, 1960:159) to suggest the abandonment of the term "Motilón" and its derivatives as unjustifiable and misleading, and to substitute the terms Dobokubí (formerly "Wild Motilones," Kunaguasayas, and "Mapes") and Yupa (formerly "Tame Motilones" and "Chakés"). Finally, however, Wilbert (1960:103) introduced the more appropriate autodenomination Barí (Barira), "people," for the Chibchan tribe, based upon a vocabulary that he (in collaboration with Villamañan) collected of their language. At the same time the Japreria (Sabril, Sápriria) subtribe, which had been excluded by Rivet and Armellada from both the Cariban- and Chibchan-

[8] In Colombia they are thought to have extended 50′ further south (Reichel-Dolmatoff, 1960:194).

speaking tribes because of lack of evidence, was included by Wilbert (1960:101) among the former, when he analyzed a basic vocabulary of their language that had been collected by Villamañan and Santelos (1956–1957).

Subtribal Identification

In addition to the bewilderment occasioned by the loose usage of the term "Motilón," the plethora of names which has been applied to subtribes and smaller groups further confuses an examination of the literature. Two basic groups are recognized among the Yukpa-Yuko: the Yukpa who inhabit Venezuela, and the Yuko who live in Colombia. This is not simply a function of the international border, but rather it represents a division recognized by the Indians themselves. The division is reinforced by physical, cultural, and dialectal differences which have probably resulted from the geographic and historical causes and from the high degree of inbreeding. These differences extend to the level of the subtribe. According to Reichel-Dolmatoff (1960:165–166) the term "Yuko" is applied by the Indians themselves to indicate "savage Indian," whereas "Yukpa" means "tame Indian." Despite important differences between the Yukpa and the Yuko, and even within the subtribes comprising them, there are sufficient similarities which permit them to be classified as one major cultural unit.

At the present time, the Yukpa are divided into eight subtribes: Irapa, Japreria, Macoita, Parirí, Shapurrú, Viakši, Wasama, and the Rionegrinos.[9] The Yuko also comprise eight subtribes: Iroka, Las Candelas,[10] Manaure, Maracá, San Genaro, Sokomba, Susa and the Yowa. Other names appear in the literature, some of which were applied to groups that have subsequently been eliminated by merger or by extinction. The majority of these additional terms, however, refer to local groups or even to individual extended families with the subtribe. In earlier times these groups may indeed have comprised distinct subtribes, but in the contemporary context the use of very localized names on a par with those of the major groups only compounds a confused situation.

Historical Perspective

Before the arrival of the Europeans a perpetual state of war existed between the Yukpa and other tribes, and among the Yukpa subtribes as

[9] The Rionegrinos refer to themselves as the *Wošípore* ("people of the savanna").

[10] These were also referred to as the Tukušmo (Reichel-Dolmatoff, 1960:162). Reichel-Dolmatoff (ibid.: 163) also mentioned the existence of the Curumaní groups, who lived northeast of the *campesino* settlement of Curumaní. These no longer exist as a separate entity.

well; for, as has been suggested by glottochronological analysis (Layrisse, Layrisse, and Wilbert, 1966:422), the subtribes had begun to diverge just before the Spanish Conquest. The Barí frequently raided the Yukpa in order to kidnap women and children,[11] and enmity existed between them until the 1950's. Further, according to the Yukpa oral history, the subtribes were also continuously at war with the *wanápsa*,[12] the original inhabitants of the region. The *wanápsa* were unable to resist the onslaught of the Yukpa as the latter moved into the mountains and were eventually annihilated.

More significant has been the persistent warfare among the subtribes, and between extended families of the same subtribe. Alcácer (1962:17), in his study of the documents relating to the history of the region, noted that during the seventeenth century the Coyamos were at war with the Aratomos and the Chakés, whereas the Aratomos were also at war with the Sabriles. The Sabriles, apparently the least tractable of the subtribes, were at war with all of the other Yukpa groups. More recently de Booy (1918*c*), Reichel-Dolmatoff (1945), Bolinder (1958), Wilbert (1960), and Ruddle (1970*a*) noted the phenomenon of intratribal warfare.

Fieldwork in 1969 indicated that strong animosity still exists among the subtribes. Among the Yuko, enmity is great between the Maracá and the Iroka; whereas in Venezuela all of the subtribes, but especially the Rionegrinos, are on bad terms with the Parirí, who themselves have recently been sundered by internal strife.

Despite the bloody legacy, alliances and peaceful marriage unions have taken place among the subtribes; and in the mission settlements of the eighteenth century (Alcácer, 1962:23), Yukpa from a number of subtribes were able to live peacefully together. De Booy (1918*b*:392) reported that a number of Tukukus (Irapa) who had quarreled with their own group were living amicably with the Macoas (Macoita); and Bolinder (1958) mentioned the friendship of the Pišíkakaw (Síkakaw), an Irapa group, with the Maracá, which continues to the present day.

Doubtless such internecine, intra-group bloodletting has had a greater effect upon the demographic, social, and even physcial attributes of the Yukpa than has the impact of *wátia* colonization in the lowlands. It is probable, however, that hostilities among the subtribes were heightened

[11] Instances were recorded in 1744 and 1769 (A.N.C., *Curas y Obispos*, t.20 f.814).

[12] "*Wanápsa*" is a Yukpa term. These may be the same people referred by de Booy (1918*b*:498, and 1918*c*:212) as "Arahuacos."

by the impact of the *wátia* (whose depredations and settlement forced the Yukpa into an increasingly inhospitable, alien, and marginal habitat) which made imperative the definition of subtribal territories for subsistence purposes. Thus warfare resulting from territorial competition reinforced that stemming from raiding for females.

The Yukpa were first exposed to the violence and wantonness of the conquistadores in the 1530's when Alfinger and his lieutenants explored the shores of Lake Maracaibo, crossed the Sierra de Perijá, and then passed southwards down the "Valle de Upar." In his frenetic search for gold, Alfinger devastated the whole area, burning settlements, massacring the Indians, and putting the survivors to flight. Some fled to the fastness of the Sierra Nevada de Santa Marta, but the majority sought refuge in the Sierra de Perijá (Simón, 1882–1892:v.5, 233).

So it was that from the earliest days of the conquest until about 1960 the pattern of recurrent violence which was to characterize the relationship between the Indians and the *wátia* in the territory between Lake Maracaibo and the César Valley was established. Some of the Yukpa later ventured into the lowlands when offered guarantees of safety and were settled on *encomiendas*. This arrangement did not endure, as they pined for the freedom of the mountains,[13] to which they returned and where they remained until the arrival of the Capuchins in the late seventeenth century. Mission stations in which Yukpa and in some instances selected *wátia* families were settled were established both in the western Maracaibo Lowland and the César Valley between 1694 and 1750, and the town of La Villa del Rosario was established in 1768 with families from the Canary Islands (Lodares, 1930:394). Unrest continued despite the efforts of the missionaries, who had to contend with bureaucratic wrangling and malfeasance,[14] especially in the Province of Maracaibo, in addition to the continued intractability of the Sabriles and those subtribes living south of the River Apón. The Indians, resentful of the *wátia* appropriation of their ancestral lands and their continual provocation, repeatedly attacked settlements, both ecclesiastical and lay, to the south of Lake Maracaibo, in the western lowlands, and in the César Valley. Punitive raids upon the Indians only increased the general level of violence, such that by 1771 the Yukpa struck repeatedly in the

[13] Archivo General de Las Indias, Seville. *Santafé*, 547.

[14] Although this undoubtedly was not all one-sided.

lowlands, forcing the abandonment of an *hacienda* in the Palmar Valley and completely cutting all communication between La Villa del Rosario and Maracaibo.[15]

Relationships between the Yukpa and the *wátia,* which deteriorated greatly during the last quarter of the eighteenth century, were completely severed in 1821, when the Capuchins were forced by the turmoil of the War of Independence to end their missionary activities. All but the most acculturated Indians withdrew once again to the mountains.

All knowledge of the Yukpa country and its inhabitants was forgotten. The transmontane trails between La Villa del Rosario and El Molino fell into disuse;[16] and only sporadic and generally violent contact took place between the races, as the *wátia* progressively usurped the Indians' former lowland territory and converted it into cattle ranches and small holdings.

Although a mission station had been established at Codazzi in 1910[17] and peaceful contacts with the Yuko had been re-initiated in 1913, Bolinder (1958:24) in 1915 found the César Valley had been depopulated by Yuko attacks and the settlements of Palmira and El Jobo sacked. Between Codazzi and Becerril, only one cattle ranch, at Latio, a short distance north of Becerril, remained.

In the Maracaibo lowlands the situation was a little better. Although the *wátia* had settled as far south as Machiques by the late nineteenth century, in 1917 hostile territory still commenced at Los Cañitos on the River Yasa, a short distance further south (de Booy, 1918*a*:487). According to de Booy

---

[15] A.N.C., *Milicias y Marina,* t.121 f.121.

[16] One was constructed in 1737 by Padre Silvestre de Lavata (A.N.C., *Conventós,* t.15 f.548; and A.N.C. *Eclesiásticos,* t.12 f.414). Simons (1881:711) mentioned that an old trail once existed, "not many years ago a rugged footpath led across the high ranges of the Andes from Molino to Perijá (La Villa del Rosario)." Sievers (1888:113) wrote that "as late as the sixties a path existed from El Molino . . . which led to the lowlands of Lake Maracaibo at the Palmar River." De Booy (1918*a*:397) learned that this path led along the River Palmar and then followed one of the tributaries (either the Tosas or Lajas) to its source, whence it descended without difficulty into Colombian territory. It is unlikely that the route constructed by Lavata followed the Palmar because of the perpetual hostility of the Sabriles who inhabited the area. More likely Lavata's trail followed another route mentioned by Sievers (ibid.:113–114) which led from Espíritu Santo (Codazzi) to Venezuela. De Booy (ibid.:387–388) stated that neither the Macoas (Macoitas) nor the inhabitants of Machiques had any knowledge or legends regarding such a trail. This is not surprising in my opinion, for, so far as the Venezuelans knew, white men had never entered the mountains from the Venezuelan side. They had evidently forgotten all knowledge of the priests' explorations.

[17] Archivo del Vicariato Apostólico de Valledupar.

(ibid.) a trail had once linked Machiques with El Pilar on the Catatumbo, but it had remained unused during the nineteenth century owing to attacks by the Parirí, Tukuku (Irapa), and Rionegrino subtribes of the Yukpa, and by the Barí.[18]

Bolinder (1958:207), who passed over the mountains from Colombia and descended to Machiques via the Negro Valley in 1937, showed that, although the Venezuelans had instituted a crash program to civilize the Yukpa, settlement was still halted at the Yasa frontier. A good mule trail had been cut from Machiques to Ajayapayna in tthe Negro Valley, and the Rionegrinos, who had received a great many gifts from the *wátia* were sufficiently acculturated to have become coarse and rude. At this time in the César Valley the Yuko had become pacified enough to have limited, peaceful intercourse with the Colombians, largely as a result of the efforts of the Capuchins, and contacts with the *campesinos,* who continued to usurp Yuko lands, no longer always led to bloodshed.

Although the peripheral location of the Yukpa country near the northwest margins of South America had insured it the early attention of the Spanish conquerors, and despite the early visits of Europeans to the region, it remained largely unknown and hostile territory until the second half of the twentieth century, its history dominated by continual inter- and intratribal warfare, and the struggle against the conquerors. Conditions began to change in 1947, when the Capuchins reopened their "civilizing" campaign among the Yukpa, with the establishment of a mission station at El Tukuku. Cattlemen, who had previously been warring with Indians (Armellada, 1946:368) followed the priests and by 1960 the *wátia* frontier had reached the River Tukuku, the former southern limits of the Yukpa territory. Happily, in Venezuela an official Indian Reserve was established at the eastern foothills of the Sierra de Perijá to prevent the *hacendados* from stealing more of the Indians' land, and subsequently, the Yukpa have been able to descend again to live at lower altitudes.

The pattern in the recent history of the Yuko, where a Capuchin mission was established at Socorpa in 1960, has been similar to that of the Yukpa.

[18] Illustrative of the legends that had been built up about the fierce Yukpa is the report by de Booy (1918*a*:497) of a semi-aboreal tribe known as the "Porotos," whose ferocity was dreaded even by the Motilones. *Poroto* is the Yukpa name for the *marimonda* (*Ateles belzebuth*).

In Colombia, however, in the absence of an official Indian reserve it has been easier for *hacendados* and *campesinos* to continue to steal Indian lands.

Contemporary Cultural Interfaces

During the course of their history the Yukpa have been harried by many of the neighboring tribes. Despite these hostilities, however, a number of cultural traits were acquired from other tribes (Metraux and Kirchhoff, 1948:350; and Layrisse, Layrisse, and Wilbert, 1960:423).

To the south lies the territory of the Barí who have frequently preyed upon the Yukpa, robbing them of women and children. To the north live the Parajuano and Guarjiro Indians, Arawakan tribes, with whom contacts in the past were, apparently, generally unfriendly. According to Uscategui (1961:218) some contact resulting in trade occurred between the northwesternmost Yuko groups and the Arawakans of the Sierra Nevada de Santa Marta.

In recent years, with the coming of peace, intertribal contacts have become friendlier. A number of Guajiros now live among the subtribes west of Machiques, and there is intermingling between the Yukpa and the Barí in the southern end of the Valley of the Motilones. Indeed, so improved are relations, that in the mission at El Tukuku a number of mixed Barí-Yukpa marriages have taken place. By far the greatest influences upon the Yukpa have come from the *wátia,* from whom the Indians are gradually obtaining all of the trappings of "civilization."

Population Size

Various estimates of the total Yukpa population have been made, but generally without any indication of the size of subtribal units. Layrisse and Wilbert (1966:51), however, estimated in a general way that the total population of each subtribe was less than 400–500 individuals. Hildebrandt (1958:92) estimated the total Yukpa population to be about 1500 individuals, and two years later the Venezuelan national census guessed that they numbered 2057. Reichel-Dolmatoff (1960:162–163) provided data by subtribal units and estimated the Yuko population to total 580.

Data which I collected in the region during 1969 and 1970 indicate a total estimated population of the tribe in Colombia and Venezuela of about 2176 persons, composed of 1484 Yukpa and 692 Yuko, divided as follows:

YUKPA:		
	Irapa	588
	Japreria	47
	Macoita	250
	Parirí	101
	Rionegrino	120
	Shaparrú	50
	Viakši	9
	Wasama	70
Tukuku Mission Village	(minus Irapa)	149
Mission school children		100
	Subtotal	1484
YUKO:		
	Iroka	370
	Las Candelas	15
	Manaure	35
	Maracá	120
	San Genaro	10
	Sokomba	30
	Susa	100
	Yowa	12
	Subtotal	692
	Total Tribe	2176

The data are most accurate for the Irapa and Maracá, which were thoroughly surveyed. Other subtribes were estimated with the help of trusted Yukpa informants, priests, and others whose information supplemented the data obtained on my brief survey visits. The figures differ from those which I published earlier (Ruddle 1970*a*:31) because of later data.

Within the Yukpa-Yuko country the population density, which averages 1.5 persons per square kilometer, varies considerably. The average number of homesteads per settlement among the Irapa and Maracá was found to be two, containing an average population of 3.3; of whom 1.3 were adults of working age.

Distribution of Population

Traditionally, each of the subtribes has occupied a distinct territory focusing upon a particular river valley (Map 2). Within these valleys the subtribal population is divided into smaller settlements, the majority of which are located upon the terraces of the major valley and some of the tributary streams. Land in the interfluves is virtually unoccupied. The settlement pattern is a mixture of single houses, hamlets composed of two or three houses, and larger nuclei containing as many as 16 houses.

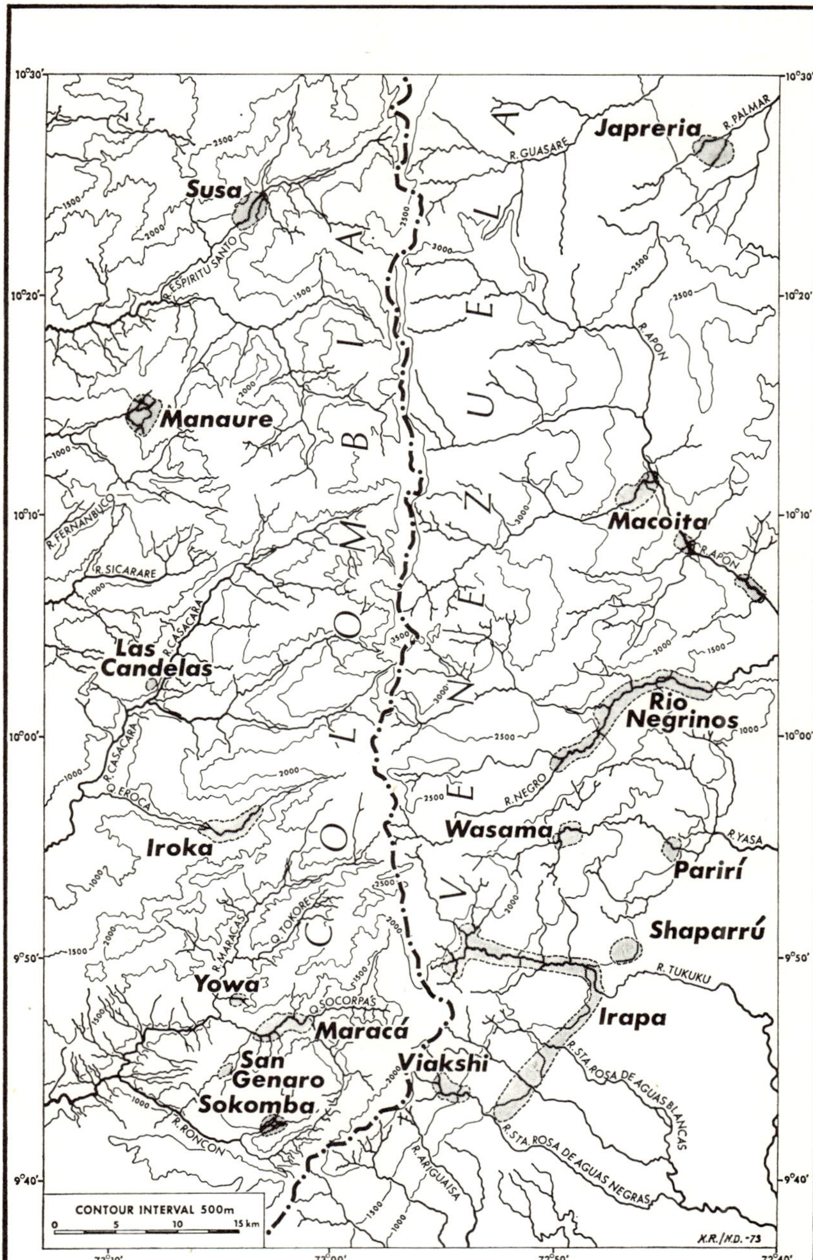

Map 2. Location of the Subtribes.

During the last quarter of a century, however, fundamental changes have modified the traditional settlement pattern of the Yukpa-Yuko, which reflected, in part, the intense hostility between the subtribes, in addition to the scarcity of subsistence resources. The increasing competition for land with Venezuelan and Colombian settlers during the last two decades, together with the pacifying influence of the Capuchin missionaries, who have been active at the Tukuku mission (Venezuela) since 1945, and at the Socorpa mission (Colombia) since 1961, has led to the regrouping of many Indians into larger settlements. At the present time the *Strassendorf,* at El Tukuku mission, is the largest settlement in the tribal lands. In that village, 415 persons from eight subtribes live harmoniously together under the close supervision of the priests.

The majority of the Irapa inhabit the middle and upper reaches of the Tukuku Valley at an elevation of from 300 m. to 2000 m. As a result of the Capuchins' most recent efforts, which have brought peace to much of the region in the last 25 years, and particularly since the pacification of the Barí at the start of the last decade, some Irapa have emerged from the interior of the Sierra de Perijá, and have settled in the lowland area known as "El Valle de los Motilones." Irapa settlement now extends southwards along this valley to the River Santa Rosa de Aguas Negras, which was formerly deep inside Barí territory.

Similarly, the Macoita, Parirí, and the Rionegrino subtribes occupy the middle courses of the Apón, Yasa and Negro (Atapši) Valleys respectively at an elevation similar to that of the Irapa. In more recent years, these tribes, too, have manifested a tendency to resettle at lower altitudes. The Shaparrú inhabit a single settlement at an elevation of 530 m., approximately two miles north of the River Tukuku in the upper reaches of the Shaparrú Valley, a tributary of the Yasa.

The Viakši and the Wasama groups live at elevations of approximately 1700 m. The Viakši, an extremely small group, inhabit the headwaters of the River Santa Rosa de Aguas Negras, whereas the Wasama live upstream of their kinfolk, the Parirí, along the River Yasa.

The situation vis-à-vis the Japreria group is somewhat confused. Their principal settlement sites are located along the headwaters of the Palmar, although recently unsuccessful attempts were made to relocate them along the Caño Pescado. In late 1969 I encountered a large segment of this group which, it appears, had temporarily taken refuge in the mountains east of Villanueva in the Colombian department of La Guajira.

The Yuko of Colombia occupy niches similar to their counterparts in Venezuela. The Yuko, however, have been forced to relocate their settlements at higher altitudes, owing to the progessive alienation of their lands by increasing numbers of Colombian smallholders during the last two decades.

The most numerous Yuko groups are the Iroka, Maracá and Susa which occupy the upper courses of the Quebrada Eroca, Quebrada Socorpas and the River Espíritu Santo respectively, between the elevations of 1000 m. and 1500 m. The Sokomba group dwells south of the Maracá, in the headwaters of the River Roncón. Occupying the savannas of San Genaro west of the Maracá is the San Genaro group; and to the northwest, the area around the *finca* La Yowa is occupied by the Yowas. Members of these two small groups have close kinship ties with the Maracá with whom they are gradually merging in response to increasing pressure by Colombian settlers upon Yuko lands. The small Las Candelas group inhabits the middle reaches of the River Casacará at an elevation of approximately 1000 m., and the Manaure, which at 1750 m. is the group inhabiting the highest altitude, dwell in the headwaters of the River Chiriaimo (also known as the River San Diego).

Migratory Movements

Yukpa life is marked by both permanent and temporary movements which are both traditional and a result of the presence of the *wátia*. Permanent migrations necessitate a movement of the homestead, whereas temporary movements require only the construction of either a temporary, or often, a secondary house. Most of the people are, or have been, involved to a certain extent in one or more types of migratory movement.

Traditionally the commonest types of temporary migratory movements result from extended hunting, fishing and fruit harvesting expeditions, and from visiting relatives who live far away. These activities generally coincide with the slack periods of the cultivation cycle. People also commonly leave communities for a variable period of time as a result of an intra-group squabble. Nowadays, resulting from the presence of the *wátia,* there is a discernible, but small-scale, temporary movement of migratory labor to the missions, *haciendas,* and *fincas.* Children are also boarded during the school term in the mission at El Tukuku.

Permanent migration traditionally results from marriage, the death of a senior family member, inter- and intra-group strife, and from the need to exploit new sites for cultivation. The adoption of *wátia* ways has led to new reasons for permanent migration. Some Yukpa live and work permanently on the farms of *wátia* (although the majority prefer to work on an occasional basis) and in the missions. Permanent labor migration has resulted in only minor changes in the population distribution, compared with the desire of the Yukpa to regroup around the missions, and the theft of former Indian lands. House site shifting, as opposed to migration, takes place for the following reasons: Infestation with insects (especially cockroaches and crickets); general disrepair; to prevent the development of altercations;[19] abandonment of houses as a result of the death of a family member;[20] and to be near cultivated fields. A permanent movement of the homestead for this last reason is rare, however, because generally secondary field houses, which double as storehouses, are constructed. Some examples of this reason for house shift were noted, especially in cases where, besides distance to the fields, one of the above reasons also provided an incentive.

Health

In terms of health, it may be said that most of the Yukpa subtribes have reached a critical phase in their history.[21] Living conditions are poor. The unpleasant effects of the cold, wet climate at high altitudes are not mitigated by the thoroughly inadequate shelter provided by the traditional Yukpa houses which are adapted to warmer, lowland climates. Food is often scarce and difficult to obtain and the Indians' addiction to large quantities of nicotine, and strong liquor, together with their recently acquired predilection for frequenting the cheaper whorehouses of the lowlands, adds to the already precarious conditions of health.

In addition, the traditional shortage of women occasioned by the marriage rules and the kinship system has led to an extraordinarily high coefficient of inbreeding among certain subtribes (Diaz-Ungría, 1971:25) which has in turn given rise to the frequent occurrence of hereditary defects, particularly dwarfism, deformation of the extremities, and dropsy. I found a high incidence of such defects among the Maracá, and similar conditions have been observed among the Iroka (Reichel-Dolmatoff and Clark, 1950:107).

[19] One of the Maracá moved from a house site which his family shared with another, because his daughter was an habitual thief.

[20] The rules regarding this vary somewhat among the subtribes.

[21] The mixed group in the neighborhood of the mission at El Tukuku and the Maracá, under the care of the mission at Socorpa, should be excepted from the statement. Medical facilities are excellent at El Tukuku, where a qualified doctor and a small nursing staff (including a Yukpa woman) operate a well-equipped clinic. In addition there is a hospital at Machiques, one hour away by the missionaries' car.

Social Structure and Local Group Organization

The Yukpa subtribes are independent, politically autonomous, largely endogamous bands, which, until recently, have lived in a state of almost perpetual hostility toward each other. The several extended families constituting a band recognize a headman,[22] whose role traditionally was most important in wartime, a priest-shaman (*tómayra*)[23] who is in charge of the cult life, and a shaman (*tuwánčo* or *tuano,* "he who knows") whose principal role is that of a medical doctor. These people form a type of upper class. Otherwise class distinctions appear to be minimal. The doctor has great status because of his or her intimate knowledge of the medicinal usage of a vast number of plants. Originally, the office of the war chief, and perhaps that of the doctor, appear to have been hereditary through the male line (Layrisse and Wilbert, 1966:48). Apart from these officials, authority is maintained within the extended family by the male elders.

Among the Irapa and Maracá, residence is often uxorilocal, with bride service involved. After a few years it becomes bilocal or neolocal. This was confirmed by Wilbert (1960:116) for the Parirí. When women are abducted from other subtribes, however, residence is virilocal, to provide reinforcement against reprisals. The residence pattern does not always adhere to the residence rules; and only rarely, it seems, are households composed solely of nuclear families, even after the period of uxorilocal residence is officially ended.

As already indicated, Yukpa society is mainly structured according to kinship rather than class. The kinship system is bifurcate-merging (Layrisse and Wilbert, 1966:108), and the cousin terminology is of the Iroquois type (that is, with siblings and parallel cousins distinguished terminologically as one class of relatives, different from that of cross-cousins). Descent is bilateral.

Reichel-Dolmatoff and Clark (1950:106) have shown that polygynous marriages may be of the sororal type, something that Wilbert (Layrisse, Layrisse, and Wilbert, 1960:423) was unable to confirm for the Parirí subtribes. Sister's daughter marriages are licit, and the levirate is also practiced. Polygyny was desirable as it conferred warrior prestige upon the man who had to defend himself against the vengeance-seeking kinsmen of the abducted woman.

[22] To a varying degree, depending on the depth of acculturation.

[23] The *tómayra* obtains and maintains his position in society by receiving songs (*homáyiḵa*) during his sleep. Immediately upon receipt of the song the *tómayra* has to sing it aloud. On the following morning he has to repeat the song. If, according to his witnesses, the repetition is accurate, the song becomes the personal possession of that *tómayra.*

Real Property

In a subsistence level society the most fundamental property rights of an individual are those pertaining to land, which is the major productive resource. Land tenure rights among the Yukpa are complicated, and range from communal forms to individual freehold. Each subtribe has a recognized territory in which members of other subtribes may not exploit any form of resource.[24] The total subtribal territory is conceived of as belonging to every member of the subtribe, with all use rights vested in the social group. Similarly, noncultivable land (savanna), *šára* trees (*Poulsenia armata*), sacred pools, and ossuary caves belong to the social groups as a whole, and their alienation by individuals is forbidden.

Individual adult males exercise unrestricted control over cultivable land, which may represent either freehold by the individual or usufruct extending in perpetuity via the male line. Notwithstanding the conceptual basis, de facto control over cultivable land among the Yukpa is outright ownership by individual adult males. It is important to note that this control is extended not only over land actually in cultivation but also over forest- and fern-fallow areas which may not have been cultivated for generations. Real property rights do not end with the return of land to fallow or the death of planted perennial trees.

Such a land tenure system appears to be atypical of shifting cultivation, where usufruct rights represent a more flexible adaptation to limited land resources. Among the Yukpa alienation of land by individuals may, perhaps, reflect past conditions when there was no need to limit property accumulation by one man. Conversely it may reflect a continually dimishing available land area, so that individuals, once having obtained desirable land (especially *nóno ķuráčašķ*) are unwilling to relinquish it. However, the process by which "outright ownership" was actually achieved in the past, and the origins of the present tenurial situation may never be clearly discerned. It would seem that, in terms of the tenurial matrix established by Spencer and Stewart (1973:541), the various Yukpa subtribes are now at different stages along the scale from "Collective Usufructuary Controls" to "Permanent Occupance Controls via Mutual Authority."

Property rights not exercised by a man or his descendants over a period of many generations are bolstered by the group's oral history, but when all

[24] Except, of course, where according to the residence rules a son-in-law from one subtribe is working for a father-in-law who is a member of another subtribe.

remembrance of rights to a particular tract are forgotten, new rights may be conferred by the chief of the subtribe, acting in concert with the elders, when petitioned by a Yukpa who needs additional or new lands. In this manner provision is made for the establishment of new property rights. When, for example, refugees seek the safety of another subtribe, and when, as has frequently happened in the last quarter century among the Yuko, groups displaced by the advancement of *wátia* colonization settle in subtribal territory, provision is made for them by an agreed assignation either of vacant land (*saríkra*), the ownership of which is not remembered; or, more frequently, by a gift of land from an individual who has more than enough for immediate and foreseeable needs.

Rights to real property can be transferred by inheritance, gift, loan, exchange loan, and rental. Inheritance constitutes the most frequent method of property transferral. When a man dies, or becomes too infirm to work the land, his property rights are transferred in equal shares to his sons. Should a man have no sons, his rights pass first to his parallel cousins and brothers, and then to his uncles. A man is under an obligation to claim sufficient land so that his sons may have a good patrimony. Women do not have property rights to land.

The second most common type of permanent transfer of real property rights is by gift. Land is generally given freely to the landless regardless of their relationship to the donor. This is an extension of the Yukpa code that a family with food will never allow another to starve. Although food is always given to the hungry, such charity is limited, and the recipient is quickly encouraged with a gift of land so that he may sustain his family by his own labor.

Finally, there are those methods of temporarily transferring property rights, most commonly by means of loans and exchange loans. An individual may have enough land of his own, but lack, for example, sufficient soil of the type necessary for a particular crop. In this case he can seek permission to clear and plant in another person's land. Land may be freely borrowed, without obligation, from sons, cousins, and uncles. These loans may be reciprocal, the parties temporarily borrowing sections of land which have characteristics different from their own. The use of land may be sought from another person, in which case it would be rented and not borrowed. Payment is made either with labor, a percentage of the harvest, or a combination of the two. Borrowed land never becomes the permanent property of the borrower.

Property disputes arise when fields are made in the land of another without first seeking permission, and because of harvest theft. Under the traditional system, the illegal usage of another's land and the theft of crops were regarded as equally serious and were punishable by death, inflicted by the wronged individual. Harvested crops in the storehouse were supernaturally protected (*vide infra* p. 94); accordingly their theft was not punished by the individual. Damage caused by escaped burns is discussed below (*vide infra* pp. 77–78).

Hunting rights are communal within subtribal territory and for hunting purposes individual property boundaries are respected only out of politeness to the owner. Although any man may hunt freely on subtribal land, it is considered polite to seek permission before entering claimed cultivable land, except of course when birds and animals are actually seen damaging the crop. Such freedom does not extend to the gathering of wild plants and insects which may be collected only in a communal territory, and from a tribesman's own property. Products of the chase remain the property of the hunter, regardless of the fact that they may have originated in the territory of another, and are shared only by members of the nuclear family and those unfortunates the family may temporarily be obliged to feed.

CHAPTER 3

THE BIOPHYSICAL ENVIRONMENT

HOW THE YUKPA categorize and perceive their natural environment is dealt with in the following chapter. This chapter describes the actual environmental factors of their habitat.

GEOLOGY AND TOPOGRAPHY

The geological composition and the topographic nature of the mountains straddling the Colombian-Venezuelan border are still only superficially known, as no thorough survey of the region has been made, and no reliable maps are available. The mountains, which attain a height of about 3800 m., form the western branch of the eastern cordillera of the Andes, and are delimited to the east by the Maracaibo lowlands and to the west by the César Valley. Although they comprise numerous subparallel ranges, trending broadly from north-northeast to south-southwest, three major ranges may be distinguished: La Serranía de los Motilones and La Serranía de Valledupar, which lie athwart the frontier; and La Sierra de Perijá, which lies entirely within Venezuela.

Lithologically the region is very heterogeneous, with rocks ranging in age from the Pre-Cambrian to the Quaternary. Superimposed upon Pre-Cambrian quartzites and schists, which form the core of the mountains and emerge, together with the granites, as the highest peaks, are sedimentaries, from Lower Devonian to Cretaceous in age. The sedimentaries were intruded during the Jurassic (Wokittel, 1957:45) and again during the Miocene (Liddle, 1946:88), by granite batholiths and laccoliths.

The Tertiary orogeny manifested itself in this region by an immense uplifting of the mountains with associated folding and faulting, along a north-northeast to south-southwest axis with lateral faulting at a high degree to the major axis. The uplift was sufficient in the Serranía de Valledupar to permit glaciers to accumulate and advance during the Pleistocene as low as 2700 m. (Wokittel, *op. cit.*). Remnant glacial features are still to be found on the highest peaks, notably on Cerro Pintado.

The Quaternary era was characterized in those ranges by the erosion, transportation, and deposition of large amounts of noncompacted material in the principal valleys, and beyond the outer flanks of the ranges. During this era also, recurrent periods of uplift resulted in the formation of three distinct sets of terraces, and in rejuvenated river courses.

Although folding has been an important formative process in the area, the major mountain building is considered to be the result of a complicated pattern of faults and subsequently the formation of horsts (Hitchcock, 1954:26). These processes are very clearly manifested in the detachment of the Sierra de Perijá from the other ranges, and its separation by the fault-controlled Apón and Guasare Valleys. Within the mountains, the rivers are deeply incised in V-shaped, terraced valleys, which are mostly fault controlled. Although subsequent streams also exhibit a tendency toward control by faults, many have cut their almost vertically walled, canyon-like valleys deep into the strata of weaker rocks. Apart from a number of mesa-like formations, composed of nearly horizontal strata, areas of level land in the region are limited to the narrow river terraces, which, as will be seen, play an important role in the Yukpa cultivation.

Climate

The region occupied by the Yukpa is one of latitudinal and altitudinal climatic transitions. The latitudinal transition is dominated by a southward increase in rainfall, whereas the altitudinal changes are manifested in greater amounts of precipitation and declining temperatures with increasing altitude. There are no standard meteorological data available for the mountains, and few for the surrounding lowlands, so that the boundaries of the climatic regions can be only tentatively indicated.

According to Freile (1968), using Thornthwaite's classification, climates along the Sierra de Perijá and Maracaibo lowland contact zone range from Hot Dry (Dd A′a: La Villa del Rosario) in the north, through Hot, Semi-Humid (C_2 and A′a: Machiques), to Hot and Humid (B_3 and A′a: El Tukuku) in the south. Furthermore, it is estimated from vegetation patterns (ibid.:90) that the climate of the lowlands adjacent to the southern Sierra de Perijá, beyond the area inhabited by Yukpa, is of the Hot, Very Humid type (B_4).

In the Sierra de Perijá and the associated mountain ranges, the vertical zonation of climates almost entirely obliterates the latitudinal pattern. This, however, can only be inferred from vegetation pattern, standardized altitude-temperature correlations, and the limited observations of the author. Four distinct vertical climatic zones may be recognized in the mountain area:

Hot zone (*T. caliente*)	(0–800 m.)	(28°C–24°C),
Temperate zone (*T. templada*)	(800–2000 m.)	(23°C–17°C),
Cold zone (*T. fría*)	(2000–3500 m.)	(16°C–10°C),
Páramos	(3500 m. and above)	(9°C– 3°C).

Each of these zones (except for the *páramos,* the extent of which is too limited to be significant) tends to be increasingly arid in the north and wetter in the south, although generally the amounts are greater than in the lowlands, owing to orographic effects.

TEMPERATURE

Temperature and humidity data for the *t. caliente* are limited to a few fragmentary observations. December until March is the coldest period, with average temperature oscillating around 28°C; and July through October are the warmest months, with average temperature of about 29°C.[1] A comparison of the data available for El Tukuku (approximately 300 m.) and Socorpa (about 1400 m.) shows a monthly average temperature difference to be of approximately 6°C.[2]

RAINFALL

Rainfall data for the area are more complete, though by no means satisfactory. Unfortunately they are limited to lowland stations. Average annual rainfall amounts increase from 1124 mm. (La Villa del Rosario) through 1555 mm. (Machiques) to 3004 mm. (at El Tukuku) in the southern part of the Yukpa area. Owing to orographic effects the precipitation amounts generally increase toward the interior of the ranges, although, locally, areas of rain shadow are important. The seasonality and character of the rainfall is similar for all of the stations. There is a marked dry season from December to March (to February at Codazzi, where March is rainier than the following months). The rainy season (the rest of the year) shows double maxima, which are more obvious for El Tukuku, the rainiest station, where April, May, June, October, and November are rainy months and July through September are drier. The wettest months at El Tukuku are May and November.

Soils

With the exception of a few localized areas the soils of the Yukpa country are of poor quality. The unfortunate combination of a delicately balanced humid tropical ecosystem (in an area of extremely broken relief, domi-

[1] Approximations for El Tukuku, based upon observations made during field work.

[2] From subjective observations this difference is thought to be even greater for "cool season" nights.

nated by steep slopes) with careless land use practices and the lackadaisical use of fire, has resulted in the extensive impoverishment of the soils. Under the Yukpa system the usual rapid leaching and erosion associated with shifting cultivation is exacerbated by the failure to keep the soil covered during the cultivation cycle,[3] and by the cultivation of extremely steep slopes. The lackadaisical burning habits and indiscriminate uses of fire other than those associated with cultivation, which in many areas never permit the return of vegetation capable of restoring soil nutrients, may be the single most deleterious ecological factor in soil formation.

In Yukpa country the bedrock over large areas comprises sandstones, quartzite, and conglomerates. These weather only very slowly to give shallow, stony, relatively infertile soils, which nevertheless support fairly luxuriant vegetation if undisturbed. They are of limited agricultural value, and when cultivated require long periods of fallow, which under the Yukpa system they rarely receive because of the repeated burning by escaped fires. They degenerate into soil types known as *vípopa.* Most of the interior of the region which is not at present in forest fallow consists of this type of soil.

Toward the outer margins of the ranges are extensive tracts of rendzinas, based on limestone, which have locally accumulated into deep pockets of rich, black soil (*nóno ķuráčašķ*). Black earth of alluvial origins is found along the margins of permanent stream courses and in the seasonally marshy areas of lower stream courses (particularly in El Valle de los Motilones). More widespread are the deeper, sandy soils (*sásare*) developed upon the river terraces and quaternary riverine sediments, which when *nóno ķuráčašķ* is not available, constitute the most desired type of soil. The majority of the fields are located upon the quarternary terraces.

Other, locally occurring soil types of limited extent are recognized by the Yukpa (see Chapter 3), but they are of little importance to the cultivation system.

Vegetation

The plant associations of the region are complicated by rapid succession of latitudinal and altitudinal transitions. The orographic effects induced by the ranges are sufficient to disrupt the simple latitude transition from "Very Dry Tropical Forest" in the north to "Humid Tropical Forest" in the south, which is present on both flanks, by interposing an altitudinal

[3] Especially after the harvest in first year fields where no perennial crop has been planted.

series of humid forest associations. Espinal and Montenegro (1963) and Ewel and Madriz (1968) hypothesized, according to the Holdridge System (1964), seven theoretical natural vegetation associations in the Yukpa country:

North of 10°N latitude,[4] the lowlands surrounding the mountains theoretically should support predominantly "Dry Tropical Forest." This is replaced south of 10°N by "Humid Tropical Forest." Both of these associations ascend the mountain flanks and penetrate the wider valleys to a height of about 900 m., where they are replaced by the lowest of the altitudinal associations. Thus, the "Dry Tropical Forest," does not currently enter significantly into Yukpa territory. In Venezuela the "Humid Tropical Forest" on the other hand, does provide the milieu for the Parirí, Shaparrú and part of the Irapa subtribes.

The "Humid Tropical Forest" in its natural state is characterized by an exuberance of vegetation which is divided into three stories and dominated by the immense buttressed trees which often attain a height of more than 40 m. and diameters in excess of 80 cm. *Cedrela fissilis, Carapa guianensis, Parkia pendula, Ochroma lagopus, Warscewiczia coccinea,* and *Scheelea* sp. are listed by Ewel and Madriz (1968:97) as indicators of this association. Forest of this type is still widespread in El Valle de los Motilones, south of the Mission at El Tukuku, where, because of the recency of Irapa occupation, only a small fraction of the area is at present dominated by low secondary growth forest (*rastrojo*) and sown pastures. North of 10°N and above 900 m. the "Dry Tropical Forest" is replaced by "Humid Premontane Forest," which extends to an altitude of about 1500 m. South of 10°N, where the rainfall amounts are greater, the association is not encountered. Instead the "Humid Tropical Forest" is replaced by "Very Humid Premontane Forest." North of 10°N the "Very Humid Premontane Forest" is encountered above the "Humid Premontane Forest."

Among the taller tree species characteristic of the "Humid Premontane Forest" are: *Ficus urbaniana, Cordia alliodora, Guarea trichilioides, Croton gossypiifolius,* and *Euphorbia cotinifolia.* The lower stories contain: *Weinmannia pinnata, Heliocarpus popayanensis, Bravaisia integerrima,* and *Miconia serrulata,* together with a large number of herb species and palms (Ginés and Foldats, 1953:327). Epiphytes, vines, and creepers are abundant.

[4] Approximately an east to west line between Machiques and Codazzi.

In general the same species also characterize the "Very Humid Premontane Forest" which extends to about 1800 m. above sea level; but in this association, species intolerant of abundant moisture and a lower temperature are not encountered. Although the division into three stories is maintained, examples of the immense trees encountered at lower altitudes are found more rarely. The understory of small trees and bushes is much denser.

At about 1800 m. this association is displaced by cloud forest vegetation of the "Very Humid Low Mountain Forest," which extends up to approximately 2600 m. The distinctive three level pattern of the vegetation is continued, but the overall stature of the trees is smaller, and the understory is denser and less penetrable. Epiphytism is omnipresent, and tree ferns and homogeneous fern associations are common.

Above 2600 m. is found "Very Humid Montane Forest," a cloud forest association. At this altitude low temperatures limit the growth of arboreal species, and the upper canopy attains a height of about 10–15 m. The trunks are very ramified and the crowns rounded, and there is a high degree of epiphytism.

Above 3500 m. the highest peaks support sub-alpine, wet *páramos* and pluvial alpine Tundra. These areas were not visited during the course of fieldwork.

Centuries of forest clearance and repeated burning of the vegetation have drastically altered the vegetational characteristics of the region inhabited by the Yukpa, so that even in the least densely settled and more heavily forested northern parts of the ranges it is extremely unlikely that any primary forests remain.[5] The vegetation of the region presents at first sight a bewildering mosaic of various grades of secondary forest, interspersed with tracts of grassland, some of which comprise only local interruptions in the forest, whereas others extend over entire mountain ranges.

Over most of the area under study, extensive tree growth is confined to gallery forests in the valleys and to the lower slopes of the ridges. Many parts of the region are dominated by grassland formations[6] of anthropogenic origin, which have developed in sites where the return of the climax vegetation has been inhibited by frequent and uncontrolled burning.

[5] Vegetation succession on fallow field sites is discussed in Chapter 4, pp. 118–121.

[6] Known locally as *"sabanas."* I have used the term "savanna" throughout this study to refer to these formations.

Although those so-called savanna formations are dominated by Gramineae they also contain extensive tracts of ferns and straggling thickets of low, fire-resistant, ligneous growth. These savanna formations occur particularly on steeper slopes, ridges, and gently sloping mesas which are repeatedly exposed to escaped or intentionally set fires. Many savanna and fern associations are found in the pauperized soil developed on sandstones, conglomerates, and quartzites which render the return of trees difficult.[7] Indeed, indiscriminate and repeated burning has upset the frail ecological balance characteristic of most of the region to the extent that trees may never be able to recolonize some areas within the foreseeable future.

Among the Gramineae most commonly encountered in the savannas are, according to Ginés and Foldats (1953:336–340): *Pennisetum bambusiforme* and *setosum, Panicum maximum, Melinis minutiflora, Hyparrhenia bracteata* and *rufa, Axonopus chrysoblepharis, Andropogon leucostachys* and *Trachypogon plumosus.* Less frequent in the grassland formations are *Paspalum heterotrichum* and *plicatulum.*

Widespread in the savanna tracts are members of the Cyperaceae, Piperaceae, Euphorbiaceae, Verbenaceae, Scrophulariaceae, Rubiaceae, and Compositae families. Both the *Bulbostylis* and *Cyperus* genera of Cyperaceae are common and are represented by two and three species respectively. Among the Piperaceae of the region *Piper aequale* and *taboganum* are typical of this association. Euphorbiaceae are very common, and five species of the *Euphorbia* genus are present in the grasslands. *Buchnera pusilla,* a low growing herb of the Scrophulariaceae family is also abundant, as is a Verbenaceae, *Lantana camara.* The commonest Rubiaceae are *Borreria capitata* and *Declieuxia fruticosa.* Annual and perennial herbs and shrubs of the vast Compositae family are also abundant. Among those most frequently encountered are; *Baccharis trinervis, Calea caracasana,* which with other species of the genus *Calea* is very common in the large tracts of fern, *Clibadium surinamense, Eupatorium* spp. (particularly *E. amygdalinum*), *Gnaphalium elegans, Melanthera aspera, Onoseris onoseroides, Orthopappus angustifolius, Pectis elongata, Pseudelephantopus spicatus, Senecio* spp., *Verbesina columbiana,* and *Wedelia fruticosa.*

Polypodiaceae comprise a major component of the savanna association. Those ferns most commonly encountered include *Polypodium aureum* and

[7] Hitchcock (1954:12) located the site of the Macoa (Macoita) village studied by de Booy in 1918. He comments, "The site of the old village, heavily wooded when he [de Booy] photographed it, is today an infertile stony ridge."

Pteris consanguinea. At the junction of the savanna with the gallery forest ferns of the *Adiantum* genus, which prefer shadier locations, become more abundant.

In addition to the inexorable increase of the area of grassland in the Yukpa habitat, a further consequence of the Indians' cultivation activities has been a localized example of what Parsons (1970) has called the "Africanization" of the tropical American flora. *Hyparrhenia rufa* (*Jaraguá*), *Melinis minutiflora* (Molasses Grass), and *Panicum maximum* (Guinea Grass) are sown on a small scale by a few Yukpa.

Jaraguá and Guinea Grass are most commonly cultivated in the *t. caliente.* Although the former is not recognized as one of the best pasture grasses, it is popular with cattlemen because it is self-seeding and spreads so easily and rapidly that it quickly becomes a dominant, supplanting the less palatable native Gramineae. Jaraguá is assisted in its competition with the native savanna grasses by periodic burning. Many small, isolated savanna areas, surrounded by tall forest fallow and far-removed from sown Jaraguá pastures are dominated by the grass.

Guinea Grass is one of the principal pasture grasses of both Colombia and Venezuela, where it thrives from sea level to an elevation of 1500 m. In addition to the excellent quality pasture that it provides, Guinea Grass has the advantage of being both drought and fire resistant and adaptable to adverse soil conditions. Since its introduction to the Yukpa region this Gramineae has proven to be a rapid colonizer of the thin and impoverished soils which are characteristic of the steeper slopes subject to repeated burning.

Molasses Grass is mostly sown by those Indians inhabiting the higher elevations of the sierras. Molasses Grass is analogous with the Guinea Grass of the *t. caliente,* in that it appears to thrive best on the thin and eroded soils of repeatedly disturbed sites, where it forms a thick and impenetrable blanket that successfully stifles all competition.

Although these grasses were brought into the region with the intention of producing relatively small cultivated pastures, the result of their introduction must be ranked as one of the major floristic changes made in the landscape of the region by cultivation. A few broadcast handfuls of "artificial" pasture grass seed has resulted in the almost total transformation of the floristic component of the savannas of the region in less than two decades. Native American grasses have been virtually eliminated by these aggressive competitors of African origin.

CHAPTER 4

CULTIVATION: THE ANNUAL CYCLE

THE PRIME OBJECTIVE of Yukpa cultivation, together with complementary activities, is the fulfillment of the family food needs. Only very recently has livestock production and the cultivation of additional land for incipient cash cropping and pasturage started to change the pattern of the annual cycle. Although the Indians are very interested is complementary subsistence activities, especially in hunting, Yukpa life is organized primarily around the annual cultivation cycle of the basic food crops.

PRELIMINARY CONSIDERATIONS

FOOD

The Yukpa are omnivorous. Meat ranges from invertebrates to cattle, and vegetables from forest fruits to cultigens. When sufficient food is available, the Indians distinguish between meals and snacks, but frequently, because of scarcity, this distinction breaks down. Meals have to be cooked, and must include one starch staple. The ideal meal consists of a staple crop with meat, but more often than not the latter is absent. However, with the increasing ownership of shotguns, products of the chase are now more commonly available, at least for the evening meal. Snacks consist of cultivated and wild fruits, which are eaten raw, and of invertebrates consumed either singly or in quantities insufficient to constitute a meat dish.

Starch-producing plants are the dietary staples. An estimatd 90 percent of the annual food consumed is derived from *míšķ* (*Zea mays:* traditionally grown varieties), *mayíša* (*Zea mays:* recently introduced varieties), *ķurántanašķ* (*Musa* sp.: plantains and bananas), *póyši* (*Manihot dulcis:* manioc), *sɨríašķ* (*Xanthosoma sagittifolium:* yautia) and *ķɨn* (*Colocasia antiquorum:* taro). Although the vegetatively reproduced crops constitute about 55 percent of the food consumed, their ritual and symbolic importance to the Yukpa is minimal. Emotionally, *míšķ* is their most important crop. It is the focus of their cultivation endeavors, a recurrent theme of their legends, the most savored food, and the basis of many of their social activities. In these contexts all of the other crops, including *mayíša,* are subordinate to *míšķ*. The forest is cut to plant *míšķ,* whereas the other crops are mere complements.

MEAL TYPES

Yukpa meals consist mainly of boiled or baked food and soups. Soups consist solely of the liquid remaining in the pot after food has been boiled.

Usually, therefore, the object is not to make a soup per se but to produce a meal of boiled food, of which the soup is a by-product. The most frequently consumed soups (*yóḳɨt soup; -íḳɨt* soup in suffix form) are made from leaves of yautia (*sowasíḳɨt*), manioc (*poyašiḳɨt*), squash (*ḳuvirašiḳɨt*), *tamáḳrašḳ, Acacia* sp. (*tamaḳrášiḳɨt*) and certain fungi. Sometimes the seeds of squash are ground up with the yautia leaves to make a soup. This is rare, however, as it is preferable to save all seeds for the following year's sowing. Leaves that are chosen for soup making are first ground up with a handstone (*yopótšḳ*) and quern (*yóčɨrɨt*) and then placed in a pot of boiling water. Fungi are placed whole in boiling water, "because they taste better that way."

Invertebrates, particularly mollusks and insects, are also used to make soups. They are simply tossed into a pot of boiling water and left for a short while. The liquid is then drained off and drunk. The solid matter is picked out with the hand and, after the indigestible shells and ectoskeletons have been discarded, the meat is eaten.

Soups are similarly made from all types of meat. The feathers or skin are removed from the meat, then the viscera are cut out and placed on a log alongside the fire to roast. They are eaten as a side dish with soup. The bird or other small animal is then boiled whole in water, for the Yukpa never cut their meat into small pieces to make soups.

Fish, which are used for soup making are also eaten baked and boiled. Immediately after they are caught, fish are cut open along the backbone and cleaned out. They are then placed on a specially constructed rack (*tápara*) over the fire, smoked, and dried. Meat of large animals is also smoked in this manner, "so that the 'worms' don't get into it." Smoked meats are placed in a basket (*menúri*) and hung from the rafters. It is claimed that meat thus preserved is edible for up to four weeks.

By far the most frequently used methods of cooking are boiling and roasting. All foods can either be boiled or roasted according to the whim of the cook. Boiling now is done in aluminum pots obtained from the *wátia* but up until a decade ago clay pots (*pášḳ*) were used. Roasting techniques are, to say the least, rudimentary. Such items as meat, maize,[1] manioc and plantains are simply placed on a hot log over flame-free embers. Recently some of the more acculturated groups have adopted the trait of placing the

[1] According to the Yukpa, maize should never be permitted to come into direct contact with flames. See also Bolinder (1958:56).

roasts on a wire mesh over the fire. Nothing is wasted in roasting. Unfortunately the results are not normally culinary masterpieces, as the techniques of the Yukpa generally produce a well burnt exterior and a raw interior to the joint.

Relative Importance of Food Crops in Order of Production

The emotional importance of cultivated foods and their importance on the scale of production and availability do not coincide. In order of production the food crops are: plantains and bananas, manioc, traditional maize, introduced maize, yautia and taro, hyacinth bean and other varieties of legume.

Míšk is the crop with the greatest symbolic importance and the one the Indians say that they like more than any other crop, even though other crops yield more and are available all year round. It would be much more difficult to obtain the same food production from *míšk*, for a much greater effort would have to be expended to clear a larger area of forest; and, in any event, there is insufficient good land available for a subsistence based entirely upon *míšk*. Furthermore, although it is the symbolic dominant, the Indians do like the variety of the other food sources, and, too, variety is an insurance against natural disasters.

Tools and Technology

Students of shifting cultivation have generally commented superficially about tool inventories and technological levels, which they have tended to regard as insignificant. When viewed by those accustomed to mechanized Western systems and to elaborately prepared Andean and Asian agricultural landscapes, the technological level of shifting cultivators may perhaps be most explicitly characterized in a negative manner. Such ethnocentric attitudes have contributed to the inadequate analyses generally received by shifting cultivation systems in the literature. The Yukpa inventory is not complex, but the documentable changes within it merit consideration, in that they provide insights into changes within the system as a whole.

Currently, as a result of the introduction of metal tools, the implements used by the Yupka are few and simple. Machetes and axes are the most important. In the last 50 years the Indians have almost completely abandoned their traditional inventory of tools and have adopted implements obtained from the *wátia*. Among the Maracá, machetes (*mašít*), steel axes (*wáša*), and *coas*[2] were accepted during the last 20 years, and currently hoes are entering into widespread use as the cultivation system evolves and weeding is included as an integral phase in the cycle.

Many of the tools traditionally employed were not used exclusively in cultivation. Axes, seed containers, collecting baskets, and drying mats were also put to other uses. The only tools with but one function were the temporary implements fabricated for one sowing or planting and then discarded.

Before machetes were obtained from the *wátia* the only tool used for field clearance was the axe. Low growth was pulled out with the hands. For the Maracá the evolution of axe technology is easily traced, because steel axes are a recent introduction. Before contacts were established with the *wátia,* a sharp stone (*tóp*), held in the hand, was used for field clearance. It was employed to girdle and pollard larger trees and inefficiently to slash lower ligneous growth. From the *tóp* developed the stone-headed axe (*víaritpo*), which comprised a sharp stone bound to a shaft of *téva* (*Guilielma* spp.) with a length of *sápara* (*Agave viridis*) fiber. The stone axe still did not permit large trees to be felled economically, so girdling and pollarding remained the standard practice. Steel tools arrived among the Maracá about 20 years ago when a number of Colombian *campesinos* were killed and their possessions, which included several fixed blade knives, were stolen. These knives were fitted to a shaft of *téva* and used in the same manner as the stone axe which they replaced. Later attacks upon newly settled *campesino* families, who were trying to establish *fincas* in the lower Roncón Valley, yielded steel axes, which induced a major and rapid change in field clearance activities. Girdling and pollarding were abandoned[3] and felling was incorporated into the cultivation system.

Steel implements had been acquired at a much earlier date by some of the Yukpa subgroups (de Booy, 1918*b*:392), and some of the Yuko, such as the Iroka (Reichel-Dolmatoff, 1945:30), and the Milagro (Bolinder, 1958:58), but these had not been passed along to the Maracá, because of the constant intergroup warfare, which almost completely precluded peaceful contacts (de Booy, 1918*b*:400, 403; Bolinder, 1958; Reichel-Dolmatoff,

[2] A digging stick fitted with a narrow, spade-like metal tip.

[3] In fact these activities were willingly abandoned, not only because of the greater efficiency of felling, but because girdled and drying trees fall without warning during windy periods.

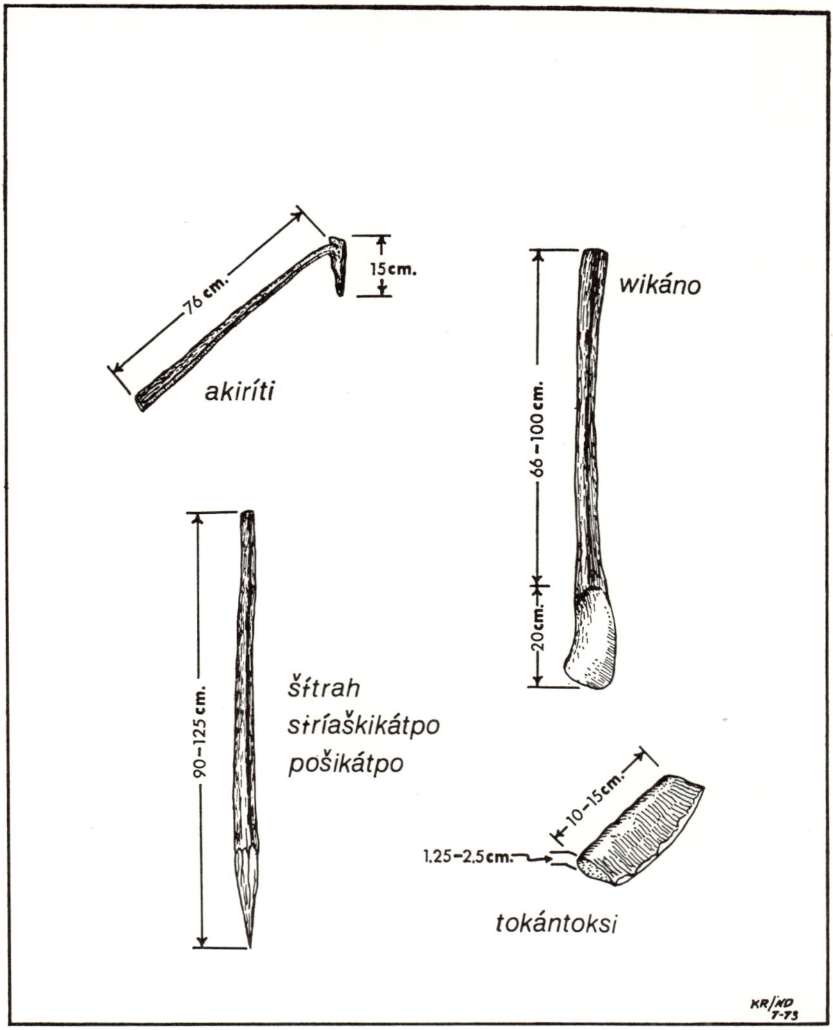

Figure 1. Traditional Yukpa cultivation tools.

1945:66; Ruddle, 1970*a*:59; and others)[4] Further, steel machetes were jeal-

[4] Some trading between the subtribes did take place. De Booy (1918*b*:392) mentions that the Tukuku Indians bartered iron mongery with the Macoa group. However, he also mentions (ibid.:400) that the Macoa were warring with the Rionegrinos. In the pattern of constantly shifting truces and alliances one pattern of hostility constantly stands out in the more recent history of the Yukpa: all of the groups seem to have been permanently at war with the Maracá.

ously guarded for making arrow heads, for use in warfare and for hunting large game. They were too precious for trade or for use in cultivation (Reichel-Dolmatoff, 1945:30).

Steel axes and machetes are now universally used by the Yukpa. Every work group has one axe for each adult male, together with several machetes and wooden grapnels (*akiríti*). The *akiríti* is used for grasping material to be cut with the machete. It is cut and prepared in the field as required. Sharpening stones (*tóp*), which are generally of sandstone, are encountered along the small watercourses. They are used *in situ,* and are never removed and taken to the house or field sites.

For sowing maize and all types of beans, the Indians cut and shape a pointed digging stick (*šíirah*); but the traditional tool for planting manioc, yautia, taro, and for crops requiring larger holes, was a separate, flattened digging stick (*wikáno*). The *wikáno* is no longer used for planting and has been replaced by the *coa,* adopted from the *wátia.* Often a machete is casually used to make holes for planting. There was no traditional weeding tool, for weeding was never practiced in the traditional system. Under the influence of the *wátia,* however, weeding is slowly assuming an integral role, and recently introduced hoes, together with machetes, are used for this new job.

With the exception of a special stone knife (*tokántoksi*) used for harvesting sugar cane, sharpened poles (*siríašk ikátpo* for yautia and taro, and *pošikátpo* for manioc), identical in design to the digging stick, were used to harvest root crops. The stems of plantains and bananas were broken with a sharpened pole (*víkan tótoma*). These tools are no longer used. Maize is harvested by pulling the cobs from the shorter plants with the hands, whereas taller plants are felled with a machete. Other harvesting equipment includes assorted baskets, drying mats (*apóte*), and sticks (*viš*) for beating beans. These implements are not used exclusively in harvesting and drying activities. Baskets are used also for transportation in general, and the mats for sleeping. With the increasing use of donkeys and mules for transportation, larger amounts can be transported at one time, so that the use of gunny sacks is beginning to eliminate the weaving of baskets. Indeed, the largest type of basket (*šámpi*), which is about the size of a sack with the capacity of 45 kilos of grain, is no longer made.

Apart from the preparation of the wet pit (*tokíru*) for taro—a task which consists of cleaning, carefully removing roots, mulching and thoroughly

TABLE 1
Cultivated Crops of the Yukpa

No.	Yukpa	English	Botanical Identification	Crop Usage 5	Incidence 6	7	Planting 8	9	10	11	Harvest 12	13
							(See General Note below)					
1	*áraš*	Kidney bean	*Phaseolus vulgaris*	Fsd	C	E	Mar. Apr.	F	D	1-1.5	Aug. Oct.	B
2	*áriš*	Pigeon pea	*Cajanus cajan*	Fsd	L	E	Apr.	H F	D	R 1.5-2	Dec. Jan.	B
3	*árros*	Rice	*Oryza sativa*	Ex	O	I	Apr.	F	D	.2	—	—
4	*čanákarašk*	Canna	*Canna edulis*	H Fss O	O	R	Mar. Apr.	F	P	1-1.5 10	?	H
5	*faráwa*	Jaraguá grass	*Hyparrhenia rufa*	G	L	I	Mar. Apr.	F	B	—	—	—
6	*ginea*	Guinea grass	*Panicum maximum*	G	C	S	Mar. Apr.	F	B	—	—	—
7	*kákašk*	Chili pepper	*Capsicum frutescens*	M-R	C	E	Mar. Apr.	P H	D [a]	R	—	T
8	*kánte*	Castor bean	*Ricinus communis*	I	O	R	W	H	P [a]	R	Sep. Oct.	B
9	*kapfwe*	Coffee	*Coffee arabica*	Cm	L	C	W	T	P	2-2.5	[f]	B
10	*kɨn*	Taro	*Colocasia antiquorum*	Fss	L	I	Mar. Apr.	F	P	1-1.6	Oct. [g]	H U
11	*kóko*	Coconut	*Cocos nucifera*	Fs	O	I	W	H	P	R	—	?
12	*kómaktašk*	Sword bean	*Phaseolus lunatus*	Fsd	L	E	Mar. Apr.	F	D	1-1.5	Aug. Oct.	B
13	*kómatarašk*	Hyacinth bean	*Dolichos lablab*	Fsd	L	E	Mar. Apr.	F H	D	1-1.5 [b]	Aug. Oct.	B
14	*kont léla*	Hibiscus	*Hibiscus abelmoschus*	M-R M	O	R	—	H F	D [a]	R	?	—

TABLE 1 – *Continued*
CULTIVATED CROPS OF THE YUKPA

No.	Yukpa	English	Botanical Identification	Crop Usage 5	Incidence 6	7	Planting 8	9	10	11	Harvest 12	13
							(See General Note below)					
15	*kurántanašk*	Plantain, banana	*Musa* sp.	Fss	U	E	Mar. Apr.	F H	P	1.3-3	Jan.- Feb. May- Jun.	H U
16	*kútia*	Amaranth	*Amaranthus cruentus*	C	O	R	Mar. Apr.	F H	B	R	[g]	U
17	*kuvišk*	Winter squash	*Cucurbita maxima*	Fss B [c]	U	E	Apr.	F H	D	R	Aug. Oct.	H U
18	*limón*	Lime	*Citrus aurantifolia*	Fs	L	S	–	H	P	–	[d]	U
19	*mayiša*	*Wátia* maize	*Zea mays*	Fss	U	E	Feb. Apr.	F	D	1-1.6	May Oct.	C B
20	*mamón*	Genip	*Melicocca bijuga*	Fs	O	I	Mar. Apr. W	H	P	R	–	U
21	*mánku*	Mango	*Mangifera indica*	Fs	O L	S	Mar. Apr. W	H	P	7-10	Apr. May	U
22	*mápaya*	Papaya	*Carica papaya*	Fs	L C	S	Mar. Apr. W	H	P	3-4	[e]	U
23	*mavú*	Cotton	*Gossypium hirsutum*	T	O	E	Apr.	F H	[a]	R	Oct.	B
24	*mišk*	Traditional varieties of maize	*Zea mays*	Fss B	U	E	Feb. Apr.	F	D	1-1.6	May Oct.	C B
25	*ñáme*	Yam	*Dioscorea alata*	Fss	C	E	Mar. Apr.	H S	P	R	Sep. Oct.	U H
26	*naránho*	Sweet orange	*Citrus sinensis*	Fs	L C	S	[f]	H	P	R	May Jul.	U

TABLE 1—*Continued*
CULTIVATED CROPS OF THE YUKPA

No.	Yukpa	English	Botanical Identification	Crop Usage 5	Incidence 6	7	Planting 8	Planting 9	Planting 10	Planting 11	Harvest 12	Harvest 13
							(See General Note below)					
27	*níspero*	Sapodilla	*Achras zapota*	Fs	O	S	[f]	H	P	R	Apr. May.	U
28	*ónoto*	Annatto	*Bixa orellana*	C	O	R	—	H	D [a]	R	Aug. Sep.	U
29	*ókote*	Balloon vine or Heart pea	*Cardiospermum halicacabum*	O T	O	R	Mar. Apr.	H F	D	R	Aug. Sep.	B
30	*pába*	Molasses grass	*Melinis minutiflora*	G	L	I	Mar. Apr.	F	B [a]	—	—	—
31	*pán*	Breadfruit	*Artocarpus altilis*	Fs	O	I	W	H	P	—	Mar.	U
32	*párus̆*	Sugar cane	*Saccharum officinarum*	B Fs	U	E	Mar. Apr.	F	P	C	[g]	U H
33	*pás̆k*	Millet	*Setaria* sp.	?	O	?	Mar. Apr.	?	D	?	?	?
34	*piña*	Pineapple	*Ananas comosus*	Fs Cm	O	C	Mar. Apr.	F H	P	1-1.6	[e]	U
35	*piriko*	Common millet	*Panicum miliaceum*	T M-R	L	E	Mar. Apr.	F H	D [a]	C	[g]	S
36	*piriñi*	Wild cane	*Gynerium sagittatum*	T	L	E	W	W	B	C	[g]	S
37	*póys̆i*	Sweet manioc	*Manihot dulcis*	Fss B	U	E	Mar. Apr.	F H	P	1.3	Jan. [g]	U H
38	*ps̆ás̆k*	Bottle gourd	*Lagenaria siceraria*	T	U	E	Apr.	F H	D	R	Aug. Sep.	S
39	*s̆ás̆k*	Sweet potato	*Ipomoea batatas*	Fss B	L	E	Mar. Apr.	F H	P	.3-1	Jul. Aug.	U H
40	*sapára*	Agave	*Agave viridis*	T	L	E	W	H	P	—	[g]	U
41	*sɨrías̆k*	Yautia	*Xanthosoma sagittifolium*	Fss	U	E	Mar. Apr.	F H	P	1-1.6	Oct. [g]	U H G

TABLE 1—*Continued*
CULTIVATED CROPS OF THE YUKPA

No.	Yukpa	English	Botanical Identification	Crop Usage 5	Incidence 6	7	Planting 8	9	10	11	Harvest 12	13
							(See General Note below)					
42	*tápašk*	Tobacco	*Nicotiana tabacum*	S	U	E	Mar. Apr.	R H	B	R	Aug. Sep.	R
43	*tomáte*	Tomato	*Lycopersicon cerasiforme*	Fs	L	I	Mar. Apr.	P H	B	R	Aug.	U
44	*tóronha*	Grapefruit	*Citrus paradisi*	Fs	O	S	[f]	H	P	R	[f]	U
45	*wáyavo*	Guava	*Psidium guajava*	Fs	O	S	W	H	P	R	[i]	U
46	*wanábano*	Sour sop	*Annona muricata*	Fs	O	I	W	H	P	R	[h]	U
47	*wɨntɨn*	Cowpea	*Vigna sinensis*	Fsd	L	R	Mar. Apr.	F [b]	D	1-1.6	Aug. Oct.	B
48	*wɨsa*	Barbasco	*Tephrosia singapou*	T	O L	E	Mar. Apr.	H F	P [a]	R	[g]	U
49	*wɨta*	Datura	*Datura candida*	M-R	O	E	May Apr.	H	P	R	[g]	U
50	*wíto*	Calabash	*Crescentia cujete*	T	U	E	W	F H	D	R	[f]	S
51	*wíto*	Bottle gourd	*Lagenaria siceraria*	*Vide supra (psásk)*								
52	*yupɨrɨ*	Mammey apple	*Mammea americana*	Fs	O	I	W	H	P	R	[f]	U

A dash indicates inapplicability.

[a] Propagated mainly by seed-fall.
[b] Also grown on trellises around houses.
[c] Seeds are sometimes toasted and eaten.
[d] Most of year.
[e] 9-14 months after planting.
[f] Very variable.
[g] As required.
[h] 3-5 years after planting.
[i] 2 years after planting.

General note:

Crop Usage (Column 5).

Fss:	*F*ood, *s*tarch *s*taple	G:	*G*razing
Fsd:	*F*ood, *s*ide *d*ish	B:	*B*everage
Fs:	*F*ood, *s*nack	I:	*I*lluminant
Ex:	*Ex*perimental	P:	*P*oison
T:	*T*echnological	D:	*D*ye
Cm:	*Com*mercial	M:	*M*edicinal
C:	*C*osmetic	O:	Body *O*rnamentation
S:	*S*moking	H:	*H*edging
M–R:	*M*agico-*R*eligious		

Incidence (*Column 6*). The crops have been rated according to the frequency and abundance of planting in fields of all ages. These terms are subjective and do not apply to the areal extent of the crop, which is dealt with in the following chapter.

O: *O*ccasional occurrence; i.e., crops either planted in few fields or planted on a small scale by all cultivators.

L: *L*imited extent; i.e., crops whose occurrence is more than occasional but less than widespread.

C: *C*ommon occurrence; i.e., crops planted by most but not all cultivators.

U: *U*niversal occurrence; i.e., crops planted by all cultivators.

Planting Data (Columns 7 through 11).

Status (Column 7). Crop plants are rated on an acceptance/importance scale.

I: Few *i*nnovators
C: *C*ash crop, with a few incipient entrepreneurs
S: Well accepted and *s*preading
R: *R*elict crop
E: Old *e*stablished, forming an integral part in the system.

Time (Column 8). This is indicated by the standard abbreviations for months. More than one notation indicates the multiple planting. The notation "W" indicates that the crop is planted whenever it becomes available. This applies especially to new plants encountered when journeying.

Location (Column 9).

F: *F*ield (any age)
H: Predominantly around the *h*ouse site
P: In field, but along *p*aths
R: Ground with *r*ocks to form terracettes
S: At base of *s*tumps in field and near house
T: Under forest *t*ree shade
W: Along *w*ater courses

Technique (Column 10).

B: *B*roadcast
D: *D*ibbled
P: *P*lanted with *coa* and machete

Spacing (Column 11). Metric distances are indicated in the table. Where these do not apply, the following terms are used:

C: *C*lumped
I: *I*nterplanted fillers, without deliberate spacing
R: *R*andom

Harvesting Data

Time (Column 12). The main harvest time for each crop is indicated by conventional abbreviations for months. Where perennials are concerned, the time of the first harvest only is indicated.

Storage Techniques (Column 13).

B: *B*askets (in house)
C: Special *C*ribs (in house)
G: *G*round
H: *H*ouse
R: Dried and *r*olled
S: *S*tacked after cleaning
T: *T*hreaded and hung
U: Direct *u*sage (i.e. not stored)

turning over the soil, and planting of a suitably humid site—no other preparation of the fields is undertaken after the burning has been completed. When the first batch of taro has been removed for transplanting, the wet pit is cleaned and dug deeply before a second batch of plants is raised in it. Fencing, corraling, and tethering prevent the entry of cattle and beasts of burden into the fields.

Cultivated Plants of the Yukpa

The diversity of cultivated plants of the Yukpa and the techniques used in the cultivation system is summarized in table 1. It includes all of the crops which are intentionally planted in the fields and around the houses. The plants have been identified and tabulated according to the vernacular crop names following the pronunciation of the Maracá subtribe. The English names most frequently associated with the plant were taken from standard sources: Bailey (1969), and Hitchcock (1951). Latin binomials are based on herbarium verification of the specimens.

Cultivation Calendar

Athough celestial movements are noted, and the Yukpa are aware of the coincidence between them and the seasonal cycle, those calendric ideas of the tribe which pertain to cultivation are organized around the passage of the seasons. Observation of the regular detailed changes in the floral and faunal components of the biophysical environment govern the timing of the annual round of autosubsistence activities. The early rains (*tiprína ántasi*)[5] arrive during April, and particularly during May, when they are accompanied by hail (*imántape*), whereas the main rainy season (*natápa*) lasts from August until the end of November. The Yukpa always know when the *natápa* has definitely set in, because savanna grasses flower. The ensuing dry season (*yéro*) completes the cycle. Its arrival is signalled by the *tátrimo* tree whose leaves turn brown and fall at this time of the year. An interval of diminished rainfall, also called *yéro*, separates the *tiprína ántasi* from the *natápa*.[6]

[5] Literally "the *tiprína* (*chichara; Cicadidae* spp.) rains." The Yukpa say "when the *tiprína* starts to 'sing' it is the signal to sprout and sow the *míšk*; when the *tiprína* stops 'singing' the *natápa* is here."

[6] Other time periods such as lunar months are identified, but they do not seem to be used with reference to the cultivation cycle. However, for the Iroka, Reichel-Dolmatoff wrote (1945:21): "Cosecha y sembrado son siempre dirigidas por la posición de la luna . . ." Yepes (1945:69) said that every month at full moon the Rionegrinos gather together to "adore" the moon, at which time they held drinking bouts lasting for two or three days.

In an examination of the timetable of Yukpa cultivation, a number of complicating factors have to be considered which preclude an orderly, compartmentalized depiction of the schedule. Wide variations, stemming from cultural causes, occur in the timing of the phases of the cultivation cycle within individual communities, and ecological factors increase the variance among communities located at substantially different altitudes. Five basic cultural factors contribute to these intra-community variations:

First, work groups have traditionally operated in an atmosphere of strict independence, devoid of any inter-group cooperation. Inevitably this has led to a personalizing of the cultivation calendar and individual needs and preferences determine variations in the time required for clearing, burning, and drying phases. The initial phase of the cycle is strenuous, and for some groups takes place at the beginning of the food-short season.[7] To supplement diminishing food supplies, and to break up the arduous routine, some cultivators spend part of their time hunting rather than continuously clearing field sites.

Second, phase timing may also be distorted by outside employment on nearby *haciendas* and in the missions. Although the Indians normally engage in such activities during the slack season of their cultivation cycle, a number of instances of such interference in the traditional routine are encountered.

Third, labor supply also constitutes a distorting factor. If the number of fit adult males in a work group is small, or if illness intervenes and additional labor recruitment proves impossible, the time required for the completion of the various phases may be greatly extended.

Fourth, sowing and planting periods may be deliberately staggered to guarantee the availability of better quality new season crops over a longer period, and to provide an insurance against the hazard of total crop destruction. Although planting intervals of this type depend upon the decision of the individual work group, a certain amount of staggering is desirable in the sowing of *míšḳ* varieties, in order to reduce cross pollination. This is a universal, however, and does not contribute to the intra- and inter-group variations being considered here.

And finally, the differential impact of innovation is also a diversifying factor. In the lower Tukuku and Negro valleys the traditional system has

[7] Although root crops yield year around and even out the supply of food, food-short periods are experienced for various reasons.

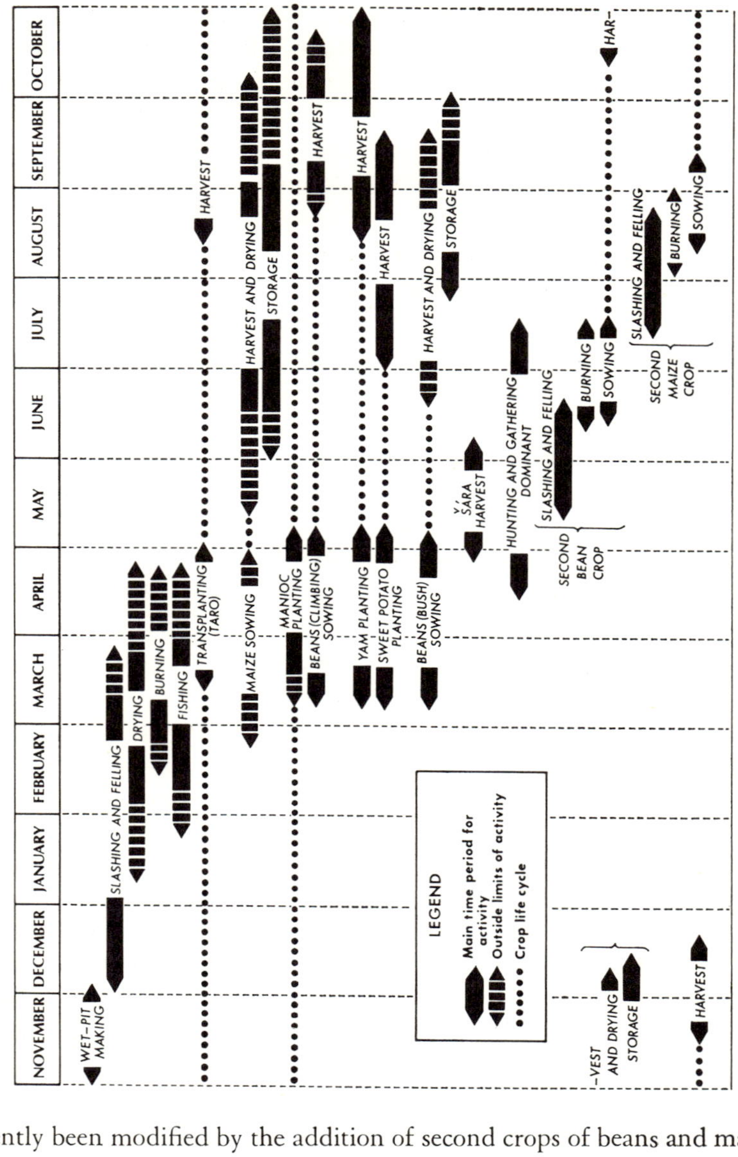

Table 2. Yukpa cultivation: A calendar for one annual cycle.

recently been modified by the addition of second crops of beans and maize, sown in separate fields and independent of the first crops. Currently, certain cultivators clear and sow fields twice in one cycle. Since these modifications to the traditional system have not yet been adopted by members of the more isolated communities, the representation in table 2 of the harvesting period for these second crops is reduced relative to that of the first crops, and allowance need not be made for the additional crop maturation time required at high altitude sites.

Intra-community variations derive, too, from a combined cultural and ecological factor. Because of the role of individual preference and the urgency of immediate needs in the assignment of priorities to subsistence activities, and as a result of the opportunities for outside employment, some work groups are not able to complete the various cultivation phases at the optimum time during the climatic cycle. This disruption of the nice coordination of the cultivation activities with the climatic regime, particularly with reference to the cutting, drying, and burning phases often results in a lengthy retardation of sowing and planting times. The net result is the complete distortion of the whole sequence of the cultivation cycle which adversely affects crop production.

Inter-community variations are more noticeable where altitudinal differences are pronounced. The important differences in the physical attributes of highland and lowland sites are reflected in temporal variations of the site clearance activities, and crop maturation requirements.

The Yukpa cultivation cycle commences in November with the selection and preparation of a wet pit for the propagation of taro.[8] November is the rainiest month, when the moisture-retentive clayey pockets along the lesser watercourses are saturated and most easily prepared for use as nursery sites.

Toward the end of November the heavy rains begin to slacken, giving way to occasional thunder showers which mark the onset of the dry season. The drenched land dries out rapidly during December and field clearance activities commence. For the majority of the cultivators, site clearance activities are well under way by the middle of December, and are usually completed by mid-February in order that advantage may be taken of the best drying weather. The numerous factors which contribute to the extension or contraction of the time required for the completion of this phase insures that in some part of the Yukpa country clearance activities are taking place from the beginning of December until the last week of March. Two cases of the late termination of clearance activities (terminating on the 27th of March) were noted in 1969 among the Shaparrú.

The timing of all other phases of the cultivation cycle hinges upon the

[8] Preplanting of taro is not practiced by all of the subgroups, or, indeed, by all of the work groups within a given subgroup. Thus, for some cultivators, the annual cycle begins in December.

completion of field clearance, and the onset of the early rains. The fields of those cultivators who complete clearance early have ample time to dry thoroughly, which, in part, ensures a good burn. Others who complete clearance activities relatively late often allow insufficient time for drying, and even miss completely the best drying weather. The drying and burning phases often overlap into the first few days of the *imántape* in the most ill-timed cases. Coincident with the drying phase is the fishing season (Ruddle, 1970*a*:276–280).

Those who burned early as a precaution against the redampening of their debris by the premature onset of the early rains are forced to wait before transplanting taro from the wet pit. By contrast, the late burners proceed directly from burning to transplanting. Transplanting operations from the wet pit to the dry hillside should not take place too far in advance of the early rains to avoid dessication and death of the young plants.

Traditionally transplanting of taro and sowing of maize coincided, the men concerning themselves with the former; and the women, for ritual reasons, with the latter. The modern deemphasis of ritual activities has resulted in the diminished importance of female labor in cultivation, and now transplanting followed by maize sowing is the commonest sequence.

The extension of the sowing and planting phase results from three factors: local differences in the arrival time of the early rains; the variations among cultivators in the timing of phase commencement; and the continued addition of plants to the field by inter-planting and filling-in long after the main sowing and planting is finished.

The net result of this assortment of sowing and planting times is that the first group to start harvesting maize may do so only a couple of weeks after the last group has finished sowing. The initial harvesting of the corn in the lower Tukuku Valley starts sporadically in early May in fields which were sown at the end of February.[9] In 1969, only ten days prior to the picking of the first ears by one group, another group had just finished sowing. Thus the time allotted for maize harvesting (table 2) covers a period of five months to allow for both a near overlap in phase timing, and for the greater maturation time required for maize at higher elevations. As can be seen in table 2, this same variability applies to all crops. Its effects on root crops is masked because they are left in the ground, and harvested periodically as required for meals.

[9] Sowing this early is rare. In one example recorded the group took a chance and depended upon the nighttime showers of the dry season to provide sufficient moisture.

Owing to this variation in phase timing, when considering the Yukpa territory as a whole, it might be suggested that there is no "slack period" in the cultivation cycle. In reality this is not so. For the majority who endeavor to coordinate their cultivation activities with the climatic regime, slack periods occur late in February and March, when fishing takes place, again during May and June, and for some groups between May and August. During the second period of diminished cultivation labor, which coincides with the hungry season, greater emphasis is placed upon hunting and gathering activities. Fortunately, at this time food shortages may be partially alleviated by the *šára* (*Poulsenia armata*) harvest, a major event in the Yukpa year (Ruddle, 1970*a*:266–267). Hunger hardly ever approaches the level of starvation, because of the continuous yield of root crops.[10]

In the lower Tukuku and Negro Valleys recent additions to the cultivation system are the second bean and corn crops. Under the influence of the *wátia* some of the Yukpa now sow beans in specially prepared fields. Field clearance and sowing lasts from mid-May until the second week of August. The crop is harvested in late November and early December. Clearance of land for the second maize crop starts in the first two weeks of July, extending until the end of August, and sowing continues until mid-September. Harvesting is done in December, at the beginning of the dry season.

Cultivation Cycle

The cultivation cycle comprises five distinct phases, each dominated by one type of activity. These principal activities serve as subheadings for the chronological analysis which follows. The phases into which the cultivation cycle is divided are: site selection; cutting and felling; burning; cropping; and fallowing.

PHASE ONE: SITE SELECTION

Each adult male owns land or has clearance rights to unoccupied land within subtribal territory. Problems of site selection with reference to the requirements of other work groups do not, therefore, occur. In Colombia, however, where the Yuko are not protected by a legally constituted Indian reserve, cognizance has to be paid to *wátia* property claims, both *de jure* and *de facto,* and to land still regarded by the Indians as subtribal property. By contrast, the Yukpa, who live on an officially established reserve, are

[10] Although cases of severe malnutrition are encountered among the Yukpa, this is not attributable to deficiencies in the subsistence system as such, but rather to sociocultural factors, especially attitudes regarding work, leisure, and sex status.

now largely free of these conflicting concepts of real property. There are, however, a number of *wátia* settled illegally on the Yukpa lands, but their numbers are not sufficient to constitute a serious threat.

Within the traditional Yukpa concepts of land tenure, the cultivator proceeds with the selection of his field sites with the following principal considerations in mind: First, the size of the area cropped, in combination with the area already under cultivation, must be sufficient to meet the food requirements of the "garden family";[11] second, consideration must be paid to the ratio of available labor and capital to the area required; third, the requirements of the intended crops have to be considered, not only in terms of the total area needed, but also in terms of the ecological requirements of each crop; and finally, these considerations must be reconciled with the ecological conditions obtaining in the land owned. The need to borrow land, or to clear unclaimed land for use as supplemental plots to fulfill the reconciliation must also be taken into account.

In attending to these basic considerations, four major variables, in addition to location, are analyzed in the detailed selection of field sites: soil types; type of vegetation; terrain conditions; and magico-religious constraints.

Soil Types

The Yukpa recognize seven important types of soil, using color, sand and clay content, general texture, compaction, and the degree of leaching as the fundamental classificatory criteria. The soil types recognized, listed in order of preference, are: *nóno k̨uráčašk̨* ("black earth"); *sásare* (sandy soil); *vípo̊pa* (very thin, sandy soils); *paráyape* (moist, clayey soils); *pirápirača* (very hard, rich and exceptionally black soils); *wayík̨u* (vivid red clay found in very small patches along stream courses); and *nóno sɨwɨswɨk̨ano* ("white earth," any deeply leached soil that is useless for cultivation).

The classificatory criteria are not sufficiently sophisticated to embrace ideas of soil genesis, hence the term *nóno k̨uráčašk̨* may be applied to any fertile, black soil, the commonest being rendzinas and young alluvials. Similarly, the term *nóno sɨwɨswɨk̨ano* is applied to any soil with a grayish-white appearance, without regard to origin.

---

[11] Cf. Allan (1967:55), the "garden family" is the kinship segment which cooperates in the cultivation of a specific group of fields and obtains its subsistence therefrom. One unit is allowed for each member of the kinship segment, regardless of age or sex, who obtains the whole of his or her food supply from these fields.

The preferred soil for cultivation is black earth. This is not widespread in Yukpa country, and is confined to limited, humid areas along watercourses, where there is periodic renewal of the alluvium, and to the zones of rendzinas, developed on limestone. Black earth is regarded as the best soil for the cultivation of *míšk*, for which purpose it is generally exclusively used, although, of course, all other crops can be grown upon it. It is considered wasteful, however, to use such good soil for other crops in first year fields. If a cultivator is lucky enough to own an area of black earth greater than is required for immediate use, it will be maintained under fallow until such time as it is needed for growing *míšk*. Different types of soil will be cleared for the cultivation of other crops.

More widespread, and therefore more important in human terms is *sásare,* a deep, sandy soil which is found principally on river terraces. It is regarded as suitable for the cultivation of all crops with the exception of *míšk*. This, however, is an idealized preference which is rarely achieved in actuality. Generally *sásare* is the best quality soil available and the planting of maize on first year fields of this soil is widespread. The major problem encountered with the use of *sásare* is that it is fairly coarse, sandy, and dries out rapidly, so that when it is worked on steep slopes erosion results.

Vípopa is an extremely shallow, sandy, stony, humus-poor soil which develops largely under ferns and repeatedly burned, low forest fallow (*rastrojo*) on areas of sandstone. It is extremely susceptible to erosion and is regarded as being of only marginal value for cultivation. Although good yields of manioc, beans, and pigeon peas are derived from it during the first year of cultivation, *vípopa* areas are used increasingly either as a supplement to the rest of the cultivated area, or in emergencies when no other land is available.

It is evident from protected sites that *vípopa* should be regarded as a culturally induced soil type in the Yukpa country, resulting from uncontrolled and frequently repeated forest burning, which over the years, has led to a degradation of the fragile vegetation cover developed upon sandstone, and therefore to a progressive impoverishment of the soil.

With the exception of *nóno sɨwɨswɨkano,* which is never cultivated, and *wayíku,* which is used exclusively for ceramics, the remaining recognized soil types are of such limited extent that they are of only minor importance in cultivation. *Paráyape,* a wet, clayey soil, found along poorly drained, minor stream courses, is used for planting sugar cane, and *pirápirača,* a very hard black soil, which is difficult to work and which is used for planting all crops with the exception of plantains and bananas, complete the list of soil types recognized by the Yukpa. These last two types are of minor extent and limited value and are not therefore actively sought after by the Indians.

Type of Vegetation

The Yukpa recognize four major vegetation associations: *tuwíča* (high forest fallow); *marípta* (low forest fallow or *rastrojo*); *wápia* (fallow associations in which fern species are dominant); and *wóši* (savanna fallow associations, in which grasses dominate).

High Forest Fallow. On a given soil type the clearance of high forest fallow for field sites is not the first choice of the cultivators. When no areas of black earth are available, however, the Yukpa prefer to make their fields in areas of high forest fallow which range in age from 15 to 20 years. Older fallow is available in the more isolated parts of the mountains, particularly in the north, where the population density is substantially less. In no part of the area under study does it appear that primary forest remains.

The Yukpa prefer to cut in high forest fallow partly because of all vegetation associations in the area this yields the most ash when burnt, and the greatest crop yield to labor input. Clearance, too, is much easier, because the bulk of the undergrowth has been shaded out by the tall growth. Because the undergrowth has been greatly reduced, however, the good bed of small slash lying beneath the larger trunks is minimized, and, as a result, fields cut in tall forest fallow sometimes burn less well than other types of material. Furthermore, the debris of tall growth often does not dry out as effectively as does that of smaller material. This may result in a poor burn, particularly if the phase timing does not coincide well with the climatic cyle.

These difficulties are considered to be of minor importance when compared with the advantages to be gained by clearing in high forest fallow. By far the most important results of cutting in this type of vegetation are that the rate of weed infestation in the resultant fields is much less than that for low forest fallow, and the ravages of crop pests are reduced.

It is considered to be more fruitful to clear fields in areas where many trees in the high forest fallow have toppled for a variety of natural causes. In such sites it is often possible to combine the advantage of a deep, dark, humus-rich soil produced by the rapid decomposition of the trunks, with the benefits derived from a large production of ashes.

Rastrojo. Frequently fields are cleared in *rastrojo* only four or five years old. Although the competition from weed increment is greater than for high forest fallow, necessitating a greater labor input for field maintenance, the humus and ash production permits the production of decent crops. In many areas, especially in the territory of the Maracá, the progressive reduction of high forest fallow located within easy reach of the pivotal settlement has forced an increase in the utilization of the *rastrojo.*

In some localities, especially the lower Tukuku Valley it is considered more advantageous to cut in five- to six-year-old fallow than in high forest, especially where the presence of *toróma* (*Heliconia bihai*) and other broad-leaved plants, together with quick-growing, light-loving species, such as *širíyi* (*Cecropia* spp.), have shaded out the stoloniferous grasses. Such old field sites are identifiable by the presence of *toróma* and *širíyi;* and the former is indicative of more humid sites in which there are to be found areas of black earth. It is not an uncommon practice among the Yukpa to fallow black earth for one or two years, and then to clear it again for a further couple of years of cultivation. This rapid cycle is said not to be overly exhaustive of this intrinsically rich type of soil. One or two years of fallow black earth for one or two years, and then to clear it again for a and to permit adequate growth for the provision of a good supply of ash.

Fern Associations. Under certain circumstances, especially in emergencies, or when the need for supplemental fields arises, clearance of fern dominated associations may take place. These areas of fern formed upon *vípopa* soils are used only for growing manioc, beans, and pigeon peas. Although these areas are regarded as marginal, they do yield good crops for one year, but because of the difficulty of fern-root eradication with simple hand tools and the problem of the rapid decline of the ephemeral soil fertility, it is not deemed worthwhile to cultivate such fields for more than one year.

Savanna Associations. Large man-made tracts of grassland occur on ridges and mesas in the Yukpa country. They are not the result of over-cultivation, but rather they are fire climaxes, derived from and maintained by the careless burning of fields, and the noncultivation uses of fire. These areas are regarded by the inhabitants as useless for cultivation, but because they play an integral part in the mixed economy of the people, they are consciously maintained by frequent reburning.

The Yukpa never attempt to cultivate the savannas because they have no efficient way of eradicating the root structure of grasses. A grass problem is frequently encountered on the edges of fields cut in gallery forest which border on savannas on their upper margins. The ingress of the neighboring grasses results in a great reduction in the yield of these segments of the fields, and creates a major cultivation problem. The Yukpa attempt merely temporary measures to retard the increment of grasses, cutting them low down with a machete until such time as the crop has been harvested and the fallow vegetation can take over the role of grass elimination. No attempt is ever made to dig out the root system.

Where fire has been more judiciously employed and not permitted to escape into the savannas, slow recolonization by low trees is sometimes in process. In many areas, however, this would seem to be nearly impossible, as regularly repeated burnings have contributed to such great impoverishment of the soils as to make it unlikely that woody growth could ever reclaim the area within the foreseeable future.

Terrain Conditions

Although it is recognized that in the cutting and burning phases of the cultivation cycle steeper slopes are advantageous, sites preferred for fields have the gentler slopes characteristic of river terraces. Such areas are of very limited extent, however, and the majority of the population is forced into using other sites. The horizontal to gently inclined river terraces, with compact *sásare* soils and the moderate intervening slopes developed on limestone with their rich rendzina soils are the most desirable localities.

A number of topographic factors render particular sites unsuitable for cultivation. They are: exposure; excessive slope; excessive rockiness; marshiness; and proneness to undercutting of terraces.

Exposure. Broad, open ridge tops and exposed mesas are never considered for cultivation. The exposure of such sites renders them unsuitable for cropping, first because of the liability of the maize, plantains and bananas to wind damage (*ɨpóta pnóya*); and, second, because of the excessive drying out of the young plants (*wíčo sik̨ánpese*) owing to high evapotranspiration rates and rapid moisture loss. Soils in such unshaded sites are extremely dry and friable.

Mesas and ridge tops in Yukpa country now generally support degraded savannas on very poor *vípopa* soils. It is my opinion that these conditions have not been caused and maintained by cultivation activities *per se,* but rather by the careless and indiscriminate use of fire in ancillary activities around past and present settlement sites.

Excessive Slope. Although the Yukpa cultivate slopes up to 80 percent gradient in certain localities, the preference is for less inclined field sites. Excessive slope renders all of the cultivational activities more arduous and hazardous, and leads to serious soil erosion. Experience has shown that certain areas are very prone to landslides and slumps (*kúna tátpo*) during the rainy season, and that the cultivator runs the risk of losing all or part of his crop as he clears them. Extremely steep slopes are cultivated, but increasingly judicious selection of field sites, partly under *wátia* influence, is reducing the incidence of crop loss owing to slides. Excessive slope is recognized to be a problem because of localized erosion around plant roots, causing toppling and an extreme tendency to gullying on exposed fields.

Excessively Rocky Ground. Excessively rocky ground (*tókru*) is not cleared because the value of resultant crops on such sites does not merit the clearance labor involved. If such land is encountered during the clearance felling phase, the cutters work around it, leaving it in forest for use as a hunting preserve. If rocky land is found to predominate in a partly cleared field site, the remainder of the proposed field is abandoned and a new supplementary field site selected and cleared. Such an occurrence is, however, rare, as in general the Yukpa are very familiar with the minutest details of their land as a consequence of their hunting and gathering activities.

Marshy Ground. In the areas now occupied by the majority of the Yukpa, marshy land (*ipíkane*) is very rarely encountered. Isolated patches do occur, however, but as there is no extreme pressure upon the available land they are not cultivated. Very small marshy areas occur along stream courses within the fields and these are used for planting plantains, bananas, and taro.

River Terraces Prone to Undercutting. Terrace sites are favored for the good quality of their soils and are used principally for corn growing. Whereas these are often planted to take advantage of the good yields which result, the Yukpa realize that this is a gamble and that one runs the risk of partial crop loss owing to cave-ins which are frequent during the rainy season (see pl. 1). Often, though by no means universally, a narrow strip of natural vegetation is left on the terraces along the stream courses as an erosion control measure.

Magico-Religious Constraints

There are two types of site which may not be cleared for cultivation activities. Certain areas of forest near ossuary caves may not be cleared because of the danger of antagonizing *okátu,* the malevolent spirits of the dead. These spirits are thought to reside in tree trunks close to the ossuaries and along trails leading to them. Actually this taboo appears to be redundant, because all of the ossuaries known to me are located in eroded sections of strata high on nearly vertical and almost inaccessible canyon walls. Caves are selected for their remoteness from the pivotal settlement, thereby affording safety from marauding spirits, and for their suitability for ossuary usage. Secondly, vegetation around pools and small lakes frequented by the rain spirit *imánta* (Okorátmɨ at Socorpa, and the pond at Kunaná, in the Negro Valley) may not be cleared for cultivation, and on no account may such bodies of water be defiled by fishing, hunting, bathing, or laundering activities.

Although these taboos are acknowledged they do not normally enter into consideration during site selection. Such areas do not lie within private property. They belong to the whole subtribe. Problems have arisen in the past owing to the introduction of alien elements, notably at Socorpa, where the Capuchin priest ordered the vegetation around Okorátmɨ cut and crops planted, in an effort to eradicate one element of Yuko superstition. The crops planted in the site did not prosper and the area was permitted to return to forest. Now, because the new pivotal settlement is to be located within 50 yards of Okorátmɨ the pool is to be fenced to prevent accidents.

Other than these two types of site there are no sacred groves which may not be cut, nor dangerous trees to be avoided, despite the fact that other evil spirits (*ikáno mánponot, tayíto,* and *yukáva*) reside in the trunks of certain trees. Such spirits generally imperil the Indians only when they are hunting in dark and remote areas of the forest, when walking alone on isolated trails, and at night-time around the houses. If men are scared of passing through such areas they will carry with them a small amount of dried red pepper (*Capsicum frutescens*), which is burned to drive away the spirits. Furthermore, men are much less prone to the machinations of these spirits than are women and children, so there is generally little spiritual danger during clearance activities.

PHASE TWO: CUTTING

In this section the techniques employed for the cutting of high forest fallow and undergrowth will be discussed.

Timing. Cutting starts at the end of November and lasts until the final days of March, or, in the case of particularly tardy work groups, until April. The majority of the cultivators time this phase of the cycle to coincide with the earliest part of the dry season in order that the cut vegetation may have sufficient time to dry out. Further, during the dry period the speed at which undergrowth can regrow is minimized, so that the quantity of troublesome vegetation returning during the interval between cutting and burning is lessened. Also during the dry season working conditions are better, and there is less likelihood of the cutters losing their balance on extremely steep and slippery slopes.

Ritual Procedures. Before starting to cut the trees *sakopátpo* is performed. A grain of *míšk* is taken to the field site and planted to insure that the earth will be content and will therefore produce a bountiful crop. Then, a couple of days before the felling of the trees begins, some *túka* (unfermented *chicha*) is prepared. A large gourd (*sorúmspa*) full of *túka* is then taken to the place where the first tree is to be felled. There the cutter solemnly pours a little over the blade of the axe (*túka yapúsipi*) to insure that it will be strong enough to cut swiftly through the trees. The first tree is then felled, after which the remainder of the *túka* is poured over the stump (*túka yipákpa sitáka*). This ceremony is important, for it is believed to make the spirits of the trees fuddled and content, so that they will not prove too resistant to felling, and above all so that they will not be angry and cause the trees to fall badly, thus endangering the lives of the cutters.

Cutting Low Forest Growth. Slashing of the undergrowth (*samokápu*) and the felling of the larger trees proceed together. A patch of low growth is cleared and then the trees in the same section are felled. Thus the cutting proceeds in a piecemeal fashion until the whole of the field site has been cleared.

A path is cut into the proposed field site at the point of easiest access, and the workers spread out from it along the bottom of the area to be cut. Slowly they slash their way upslope, grasping the material to be cut with an *akiríti*, and cutting it with a machete. The slash (*ámuk*) is cut to fall downslope in an overlapping manner so that it will form an even "bed" (*nɨkókta*) of tinder upon which the ligneous material will then lie. Although a well formed *nɨkókta* is regarded as fundamental to a good burn, the slash does not generally fall evenly, and no subsequent effort is made to spread it to provide both a good "bed" and to protect the exposed soil from the elements.

During the slashing operations the workers cut lianes (*mimitakrɨnkas*) which could later cause dangerous entanglements and kickbacks when the trees are felled. Prior cutting gives them time to dry out so that they will snap more easily when the trees fall. When a sufficiently large area has been slashed the cutters return to the bottom of the slope and start to fell the larger growth.

Felling of Tall Growth. The workers proceed upslope, felling (*wášakan tomákare*) so that the trees will fall downslope (*kamá ántase*) with their crowns overlapping (*otópay iránowe*). This results in a more even burn in which the thicker logs are almost entirely consumed. Theoretically no gaps should remain to break the continuity of the cover, so that a continuous supply of small branches is available to fuel the updraft. But frequently, in practice, this is not achieved, especially since no subsequent leveling is done.

The workers fell cautiously upslope, examining the situation and condition of each tree before starting to cut it. They are especially careful to scan the upper branches in order to foresee potential snaps and entanglements (*oyánonoke*). Caution is the watchword during the cutting phase, and no informant could recall a case of serious injury or death resulting from mishaps during felling.

Felling Techniques. The principle of least effort is always strictly applied during felling. Boles are always cut above the slightest suspicion of a buttress (*yíkorat*). Nobody ever bends to cut close to the ground, therefore trees without buttresses are cut at a height of about one meter.

Trees without buttresses are felled simply (fig. 2*a*), whereas those having a large buttress are felled by a technique known as *tátrisi* (fig. 2*b*), which necessitates the construction of two scaffolds (*tapóymotos*), one on either side of the tree. This rickety construction of previously cut poles and vines is lashed together to form a rudimentary ladder and platform (pl. 2).

Further, when trees are sufficiently close together, labor-saving multiple-felling techniques (*viartíkas*) are used. When multiple felling is used, a combination of the above techniques is employed according to the size

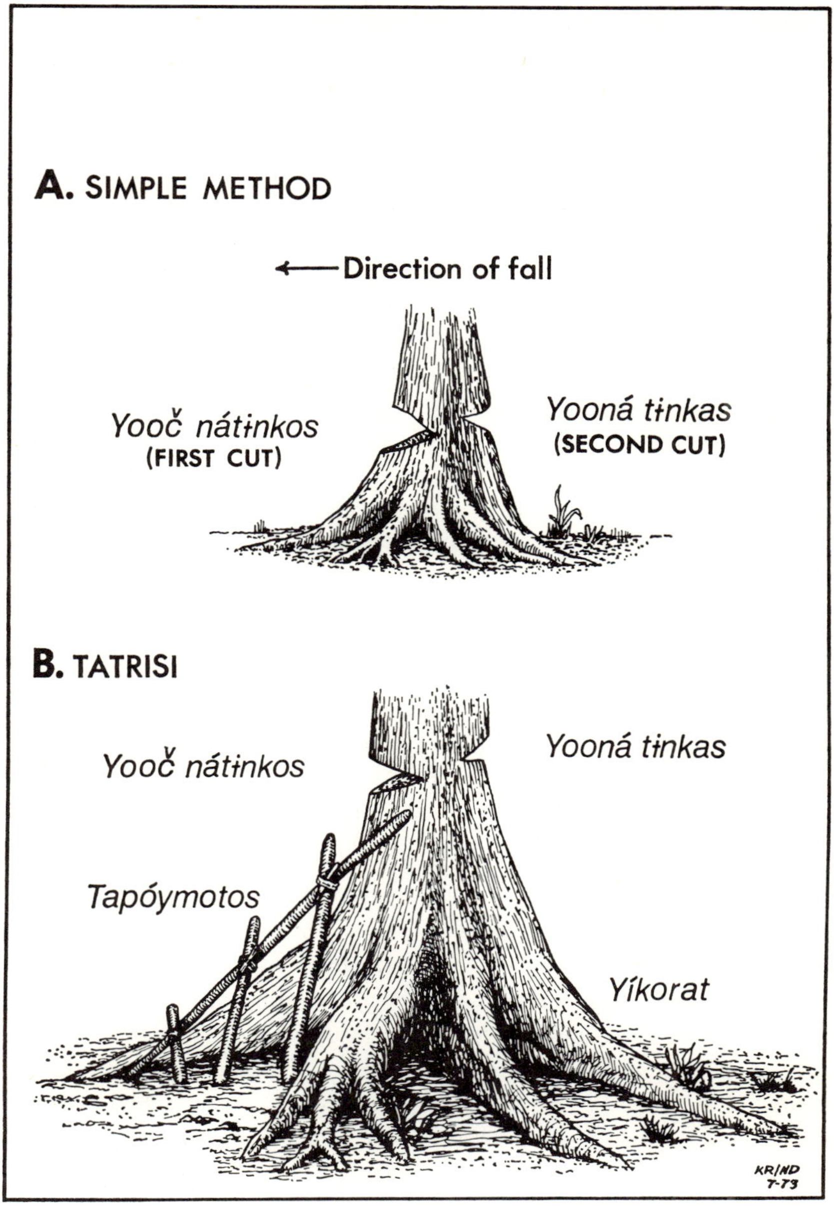

Figure 2. Tree-felling techniques.

and character of the trees involved. A large tree standing furthest upslope in the group to be felled is selected as the "king" tree. The smaller trees downslope of it are cut partly through, and then, finally, the "king" is felled completely. The force of its fall knocks the other trees down like skittles.

Mishaps. The Indians are cautious workers, and they are always careful to assess the condition and situation of each tree before starting to fell it. Yet, despite precautions, mishaps do occur. The most frequent problem is that posed by the premature snapping (*nánta*) of diseased trunks during cutting. This cannot always be foreseen, and it does constitute a hazard for the worker. Entanglements occur less frequently. They usually happen when a tree falls improperly, so that its crown becomes entangled in the branches of other trees standing to the side of the intended fall path. Dealing with entanglements is dangerous work. The cutters first carefully plan and clear their escape route. Then they proceed to cut upslope, starting with the tree furthest from the entangled trunk. The greatest hazard is encountered when trees bearing the full weight of the entanglement are felled. Finally, when the snagged trunk starts to fall free, the workers flee, dropping their axes, and running pell-mell along the pre-planned escape route.

As has already been discussed[12] relatively recent changes in the Yukpa tool inventory have brought about important changes in the techniques used for cutting vegetation. Suffice it to repeat here for the sake of emphasis that with the adoption of a steel axe and machetes the felling of trees has completely displaced the techniques of girdling and lopping.

Results of Cutting. Slashing and felling results in a *síčkápɨ,* an area of drying vegetation covering the future field. Within this area of cut and drying vegetation are to be found certain trees and other useful plants which have been spared the axe. Among the commonly occurring larger plants which are not cut down during field clearance are the *kurúkmatoro* palm (*Scheelea* sp.) which provides an edible fruit, the *téva* palm (*Guilielma* sp.), used in the manufacture of bows, and shade trees in fields in which coffee is to be sown. Other technologically useful plants that are preserved include the *sapára* (*Agave viridis*), used for rope and string making, the *nírpɨrtɨ* (*Furcraea gigantea*) which is used for the fabrication

[12] *Vide supra,* under "Tools and Technology," p. 47.

of quivers, *puuk̦* (*Stromanthe lutea*) whose leaves furnish material for weaving, and other domesticates and escapes found in *rastrojo* vegetation. Some are second degree semidomesticates,[13] which, although they are not intentionally harmed during the cutting activities, are in no way protected, and so they often suffer severe damage or death during the burning phase of activities.

The final cutting in the field site signals the end of the most arduous phase of the cultivation cycle. It is marked by all of the able-bodied men of the community passing from field to field to help finish cutting the last patches of timber.[14] This done, they celebrate with a small party (*wiyɨyak̦átpo*) for which *sóña*[15] is prepared. During the celebration the men alone dance to celebrate the end of the cutting phase. Each has an axe suspended from a tumpline hanging down his back.

The cut area is left to dry, and the community can enjoy a slack period. It is at this time of the year that fishing parties leave on extended trips and the people relax as they unhurriedly prepare for the next phase of the cultivation cycle.

PHASE THREE: BURNING

Preliminaries: Desired Effects of Burning

Good cutting and a well-burned field are inseparable. The fundamental importance of these phases is completely understood by the Yukpa. Good slashing and cutting practices are basic to a good burn, but they alone cannot insure it. Weather, timing, and dryness of the felled vegetation (*wíta*) play important and interlocking roles. Everybody is fully aware of the connection between the quality of the burn and the quality of the harvest. They recognize that apart from clearing the field of debris, preventing the rapid regrowth of weeds, and reducing the cover for animal pests, the paramount criterion for judging the quality of the burn is whether the crops are severely attacked by a variety of "worms," leaf-cutting ants (*Atta* sp.), and seed-storing ants. If the field has been properly burned losses to these pests will be minimized. To the Yukpa a well-burned field is not one in which the *wíta* has been almost entirely consumed, but rather, one with abundant ash (*viruúsa*) which is toxic to insect pests.[16]

---

[13] See full explanation of this and allied terms in Conklin (1957:44).

[14] This is the only time during the entire cultivation cycle when communal labor is undertaken.

[15] Fermented *chicha.*

[12] *Vide supra,* under "Tools and Technology," p. 47.

The Indians note too that it is much easier to sow and plant after burning. Not only is the field cleared of much of the cut vegetation, but firing also makes the topsoil more friable. In certain localities, however, especially where *vípopa* is the predominant soil type, burning results in a vast increase in the erosional potential.

Burning is accomplished in two stages: primary firing (*súrikasa*); and reburning (*tómark surikas*). In recent years, under the influence of the *wátia,* some of the Indians have added a stage of pre-burn activities and protective procedures undertaken before the primary firing.

After cutting the field, the debris was traditionally permitted to "dry out" for three or four days and then was burned. No rearrangement or leveling was done and no construction timber or firewood was ever removed from the site. Wood for any use is obtained from areas in fallow, and places where there is plenty of timber killed by escaped burns. The Yukpa say that they used to live in fear of an early start to the rainy season, which would dampen the field site beyond all hope of a successful burn. Apparently they were not able to distinguish with certainty the pre-rainy season rains from the onset of the rainy season proper. Hence they would burn directly after the *wiyɨyakátpo*. Such burns had very poor results, as the *wíta* was always too green. Planting was done on the first rainy day after the burn, whether or not this was the start of the rainy season, or just an occasional wet day during the dry season. When the full rainy season then failed to emerge for another few weeks the seedlings and cuttings dried out and died. Because of a taboo against asking for certain seeds and a general shortage of clones the Indians were not able to obtain replacements until the following season and so a food-short year would result.

Now, with the increasing acceptance of *wátia* methods, the system is changing somewhat. Pre-burn activities, protective procedures, and an increased time interval between cutting and burning are now stressed.[17]

Pre-Burn Activities

After cutting, the field is left completely alone. Everybody is extremely careful to insure that the debris is not accidentally burned prior to the time of optimum combustibility. The best insurance is to avoid the area and the drying site is not entered.

[17] Interestingly, some *wátia* shifting cultivators in the Serranía de Valledupar and Serranía de los Motilones, many of whom are recent immigrants from other parts of Colombia, are not yet attuned to the local climatic regime. They, too, have suffered total crop losses resulting from misjudgment of optimum burning and planting times.

After a predominantly dry spell of three to four weeks, during which there may be occasional short periods of rain, the cut material is examined to see if it is dry enough to produce a good burn. The condition of the underside of larger trunks is particularly studied, and they are cut into to see how well the interior has dried out. It is considered vital that these logs be dry, whereas it is thought to be relatively unimportant on the day of the burn that the material of the lighter slash (*nɨk̨óta*) may have been dampened by overnight showers, because such leafy debris rapidly dries out under the hot sun. Burning commences when the larger material is judged to be sufficiently dry.

Protective Procedures

Because of outside influences, the Yukpa have recently been made aware of the desirability of vegetation conservation. Many now cut a firebreak (*yipák̨usumik̨u*) around their field prior to burning. The cut material from the firebreak is thrown into the field, thus creating a cleared area between it and the uncut forest. However, because the firebreak is generally only one or two meters wide, it alone is insufficient to prevent the escape of fires into the forest reserve. Therefore the technique of backfiring has been introduced, which in combination with the firebreak is now serving to reduce forest destruction (*titrano*) by escaped burns (*towára wɨtɨ oyáya*). Nevertheless, escaped burns still commonly occur, resulting in widespread damage.

Ritual

Shortly before burning two important rituals are observed. The head of the garden family takes a handful of beans and a feather (from any species of bird) and throws them over the area to be burned. This is done to please the benevolent spirit of the earth (*ik̨anmɨrák̨tay*), to whom the feather smells like meat. The smell of meat in combination with the beans persuades the spirit to bring forth a good crop. Also before the burning starts the women who are to sow the *míšk̨* perform *tɨ́nis opóčpo,* in which grains of *míšk̨* are chewed and spat out onto the forearm to give it strength enough for sowing.

Taboo

Among the Yukpa an individual cultivator cannot burn without reference to the state of readiness of the other members of the subtribe.[18] Where-

[18] As in the case with so much of the Yukpa magico-religious system, this taboo is now largely ignored except in the remotest communities, where the Roman Catholic missionaries have not yet succeeded in completely subverting the Indians' beliefs.

as field cutting is an independent activity of the individual work group, magico-religious beliefs dictate that burning be a communal activity, not in labor, but in timing. According to the Yukpa system of beliefs, the smoke of the burning field causes rain. When *imánta* sees the smoke he knows that the people need rain, because they are signaling that they are ready to plant by burning the field. Rain must be sent so that the new crops will thrive. Thus, if one man burns in advance of the others there is a grave danger that the rains will render the fields of others unburnable.

Before the burning starts the heads of the work groups gather to consult with each other and to consider the state of dryness of each field. Together they decide upon a date for burning, and their decision is strictly enforced upon all members of the subtribe by the threat of violence.

One further taboo (*oyúwirpamasi*) is involved in the burning phase. It prevents the cutter of the field from also being the burner (*ámor sapóma*). It is said that if he were also to perform the burn the results would be bad, and therefore the ensuing harvest would be poor to non-existent. An uncle or nephew is generally called upon to act as burner, providing that he is skilled at the task. However, in the unlikely event that all the uncles or nephews are unskilled, one may ask other persons to perform the burn. Return of the service constitutes payment.

Only one man fires the field. Before he does so, he makes little piles of dried leaves (*tókpa*) at intervals along the lowest part of the slope. Fire is never set along the sides, top,[19] or within the field, or beneath individual trunks. Formerly the piles were ignited with a torch made of castor beans (*Ricinus communis*) threaded on a thin stick. Today, matches are used.

A good burn depends upon the dryness of the *wíta*. An ideal day for burning should be hot and sunny without much cloud, and with a moderate, but not too strong, breeze. The Yukpa prefer to burn during the warmest hours, between one and two in the afternoon, but in such broken terrain, and with tall surrounding vegetation, it is not desirable for everyone to burn at this time of day. The detailed aspect of each field determines the local timing of optimum conditions for burning. Thus, on the appointed day the hours of burning range from about 10 a.m. to 4 p.m., depending on the specific characteristics of each site.

Firing takes place in March and April (and occasionally earlier these days; see table 2) before the onset of the rainy season. Under the in-

[19] This is now changing under the modernizing influence of the priest at Socorpa, where backfiring is now encouraged, in an effort to reduce the incidence of escaped fires.

fluence of the *wátia* the cultivation system of Yukpa is in the process of change, and increasingly individuals may now burn at will, without considering the communal consequences of their actions. Additional burning now takes place in July, for the planting of the separate bean field, and again in August and September for the second crop of maize. These later burn periods are loosely adjusted to take advantage of the predictable periods of *yéro* within the rainy season.

Secondary Firing Techniques

A field in which the logs do not burn is known as a *tɨ́šrane*. This occurs when the logs are insufficiently dry to burn properly, or when the amount of leafy slash is not enough to provide tinder for the logs to catch. In these cases only the leafy matter is consumed, and the heavier timber remains. The trunks are then cut up and piled at intervals into small mounds (*túķsi*) for reburning. The *túķsi* are permitted to dry for a further short period of time, after which three more attempts are made to burn the *tɨ́šrane*. If it still remains poorly burned it is divided into two parts. In one section, known as the *tɨ́ra,* maize is planted between the *túķsi*. The other part, called the *táyñapra,* is permitted to return to secondary growth. For the remainder of the cultivation cycle the *tɨ́ra* is treated as a normal field in cultivation, while the *táyñapra* is regarded as an area in fallow to be cleared in the following year. It is considered that the addition of a further year's regrowth to the previously cut growth provides enough material for a good burn in the succeeding season.

The Yukpa appear, to an outsider, to be extremely careless of the natural environment. Until the last decade no effort was made to prevent burns from escaping into the surrounding forest. Nobody was censured or forced to make payments if his burn escaped and destroyed many hectares of forest fallow belonging to another man. Historically there had always been sufficient vacant land unowned by individuals, but belonging to the subtribe, which could be claimed. If such land was too far away, a Yukpa could always ask to borrow another's land for several seasons. Such petitions for a loan are always granted, and there are very frequent examples of outright gifts of land to the needy. At any event, such unintentional burns may benefit the owner of the destroyed forest. If perchance the area was well burned, and only large trunks remain, the owner will take advantage of the opportunity and plant the area (*owáya tɨ́ni*) in maize, squash, gourds, and beans, obtaining an additional harvest for little extra labor. He is not obliged to share his windfall with the person who accidentally burned his land.

Until recent times, therefore, the destruction of the fallow timber has not been taken seriously. However, the gravity of the situation for certain subtribes, particularly the Maracá, is now realized. Competition for land by Colombian *campesinos* has greatly reduced the area available for Indian settlement. Further, the repeated unintentional burning of the same patches of high forest has resulted in a drastic pauperization of the soils, seed supply, and vegetation, to the point where tree growth cannot compete with the infestation of fern (*wápia*) and savanna grasses (*wóši*).

One type of burn escape does, however, place an obligation on the burner. When two (or more) cultivators have adjacent unburned fields, and the owner of one decides to burn his field when the other person is absent, he is obliged to watch over his neighbor's field. Should his burn escape into part of the adjacent field he is obliged quickly and skillfully to apply fire to the rest of it in the prescribed manner. Were this not done the field of the absentee might be poorly burned and the effect would be the partial or total loss of the field. A field damaged or destroyed in this manner is known as a *towápi tini*. Causing a *towápi tíni* is grounds for killing. If, however, the burner succeeds in causing an excellent burn in his neighbor's *sə́čkapi,* the latter will be well pleased, and is obliged to make some form of repayment.

Results of Burning

For the majority of cultivators, the real results of burning do not coincide with the desired effects. For those who have slashed and felled with care, and are able to take advantage of the best time for burning, the results are generally good. But for the majority, who cut the vegetation in a haphazard manner, and who do not allow sufficient time for drying before trying to burn, the results are far from satisfactory, even after tertiary burning has been completed. In many of the fields examined the dry leafy slash had been totally consumed by the flames, whereas on small branches only one centimeter in diameter the bark had been only slightly charred. In others, concentrated efforts had been made to burn the larger trunks. These were often largely consumed, as was some of the smaller slash; but within 50 centimeters of well-consumed logs, small twigs and leafy matter remained unburned (pl. 3). There is, of course, great variation between the fields of different cultivators, and even within the same field. A major factor in ensuring good burns appears to be the degree of slope (pl. 4). On flat to gently sloping areas the bulk of the debris remains unburned, whereas on moderate to steep slopes the bulk is consumed. On the steeper slopes, however, when fire is set at the bottom to burn upwards, massive damage to the surrounding fallow vegetation generally results.

A large quantity of unburned debris remaining in the field renders the labor of the ensuing phases of the cultivation cycle more arduous, and, in addition, represents a waste of slashing and felling labor and a loss of plantable area. My observations in first year fields on flat to gently sloping (8 percent gradient) land in the lower Tukuku Valley showed that the plantable area of the field was reduced by 15 percent. This resulted, therefore, in a similar percentage of waste in the clearance labor input on first year fields; although, in the longer run, it may have resulted in the overall efficiency of second year fields being higher because there was more burnable debris. In comparison, at Socorpa, on fields with a 60–75 percent gradient, the productive area of the field was reduced by only 4 percent as a result of the debris remaining after the burn was completed. However, the damage to surrounding areas has also to be considered. Around the poorly burned fields at Shaparrú there was no major damage to the surrounding fallow, apart from some slightly scorched foliage, whereas at Socorpa, trees of the high forest fallow covering an area at least equal to that of the cleared field were totally destroyed (pl. 5*a*).

To this damage resulting from the escape of field clearance burns must be added that stemming from the non-cultivation uses of fire (Ruddle, 1970*a*:273 and 1970*b*:42–60) (pl. 5*b*). The Yukpa have no system for fighting fires. Indeed there is no desire to fight them, for to the Indians flames dancing uncontrolled across open ridgetops are an exhilarating sight. Savanna fires are mainly set for practical purposes, such as ridding an area of snakes and clearing paths, but, as has been confirmed by other authorities (Schön and Jam, 1953*b*:72), they are also kindled for their aesthetic appeal.

PHASE FOUR: CROPPING

In this section the techniques of seed and clone selection, sowing, planting, care, and harvesting of the crops is described in detail for one annual cycle only.

The Selection and Handling of Seeds and Clones

With the exception of *míšk*, the seed and clone selection procedures practised by the Yukpa show little sophistication. Although the seeds of the different crops are stored separately, and those of squash and gourds (or calabashes in the *t. caliente*) selected for taste and size respectively, only for *míšk̹* are the seeds very carefully selected and treated.

GRAMINEAE:

Saccharum officinarum. Cuttings about 30–45 centimeters long, with closely set nodes and several buds are chosen from younger cane for planting.

Zea mays. For seed corn (*míšk̹ inúš*) large, full ears, free from *k̹uríka* (a generic term applied to corn "worms" such as *Mocis repanda F.*) and Coleoptera, of the Curculionidae family (*yák̹a*) are selected. The husk (*mák̹ šúrus*) is pulled back and knotted, and the bundle is hung from the house rafters above the cooking fire so that it will be smoked (*viraskopák̹os*) and thus the damage done by the corn borers will be lessened. Traditionally, ears of the different varieties were segregated.

Shortly before planting, the ears are taken down and the tip (*yuwašač*) is cut off; its small grains are reserved for use in the *tínis opóčpo* ceremony, which takes place just prior to burning. The grains are made into *sóña* and small *k̹osíš*.[20] The *k̹osíš* are set out on the ground in front of the house as an offering to the sun spirit (*atánto*) so that he will aid in the production of a good crop.

Grains from the thicker part of the ear (*šámpasa*) are used for sowing. They are segregated according to color, shelled and placed for one day in a gourd or calabash (*wíto*) filled with cold water, so that they will sprout (*míšk̹ timási*). The day before the sowing is to start they are removed from the water and placed in the sowing basket (*mišayñatpo*). Only *míšk̹* seed is sprouted, and *mayíša,* varieties of maize recently introduced by the *wátia,* are sown unsprouted. Traditionally, the handling of the *míšk̹* seed was exclusively the work of the womenfolk.

It is forbidden to ask another person for the seed of *míšk̹*. The individual work group is expected to be self-reliant in the provision of enough seed to satisfy family needs, for it is believed that seeds given to others will grow into big plants, yielding good crops, whereas those of the donor will not thrive. When asked for seed the Yukpa become abusive, and angrily refuse the request; and it is not uncommon, therefore, to discard sections of a field which have been cleared and burned, but remain unsown for want of sufficient seed. There is, however, no taboo on the begging of food and the Indians cheerfully share their meals with the luckless, so that members of seedless work groups do not go hungry throughout the year. Surplus *míšk̹* seed is generally eaten nowadays, although in times past it was always made into *sóña.*

Panicum miliaceum. There are two methods used in the handling of seed from this plant, depending on the size of stems required. Some seeds are cut when ripe, and are hung in bunches above the cooling fire in order that the smoke will "make the seeds and the resultant arrow shafts stronger." The seeds are not harvested for selection purposes, but rather so that they may be sown sufficiently far apart to yield thick stems. When thinner arrow shafts

[20] *Kosíš* (called *hayaquitas* by the *criollos*) consist of ground maize wrapped and boiled in the husk, conserving the original shape of the cob.

are required the ripe seed is permitted to fall from the plant. The resulting plants will grow very close together, and thus their stems will be spindly and suitable for lightweight bird-hunting arrows (Ruddle, 1970*b*:27–31).

Gynerium sagittatum. Seeds are collected from wild plants wherever they are encountered. They are planted directly, "so that they will not die," in *rastrojo* alongside watercourses.

ARACEAE:

Xanthosoma sagittifolium. The cormel (*yanúš*) of the parent plant is used. It is cut off, peeled around the base and also cut into a square shape before being planted. When plants are harvested throughout the year, the clones are replanted directly in the same hole from which the harvested plant was removed a short time before.

Colocasia antiquorum. To provide clones for the wet pit the top of the corm (*yimáror*) is used. The *yimáror* is cut off, and partly peeled around the base before planting in the wet pit. Harvesting and replanting is the same as for *Xanthosoma sagittifolium*.

DIOSCOREACEAE:

Dioscorea alata. Any tuber (*yunápčašḳ*) is used for planting.

MUSACEAE:

Musa spp. Suckers (*yémoḳor*) are cut from the parent and are planted apart. Any sucker is good for planting.

CANNACEAE:

Canna edulis. Any sucker (*čánaḳarašḳ moḳóšḳ*) growing alongside the parent plant is regarded as suitable for planting. The plant is only rarely propagated from seed, as the Yukpa prefer to save the seeds for necklace making. The suckers are cut from the parent, and, if desired for crop usage, are planted about 1.75 m. apart in the fields. Recently, the Yukpa have adopted from the *wátia* the habit of using cannas as a decorative hedge plant. In this case the suckers are planted about 25 cm. apart.

LEGUMINOSAE:

Phaseolus spp., *Dolichos lablab*, and *Vigna sinensis*. Seeds are not selected from the harvest when it is being dried and stored. Later, as the supply of legumes gradually diminishes, a sufficient quantity is removed from the food store and saved for sowing. No selection criteria are applied to legumes, because "any 'bean' seed will grow as well as another." Seeds of the different bean types are separated for storage.

Cajanus cajan. Any big seed is saved from the previous harvest for sowing in new sites. According to the Yukpa, only big seeds of this plant give large, abundantly yielding trees.

EUPHORBIACEAE:

Ricinus communis. No selection is practiced. Any seed is saved for later sowing.

Manihot dulcis. Only sweet varieties of manioc are grown by the Yukpa. Cuttings (*póyšiwor*) are chosen from stems upon which buds are no more than 5 cm. apart. Those with wider spacing are not used, because "they will not give sufficient roots for the plant to grow." Cuttings are selected from the lower and middle portions near the base of the plant.

MALVACEAE:

Gossypium hirsutum. No selection is practiced. Any seed is saved for later sowing.

CONVOLVULACEAE:

Ipomea batatas. This is propagated by planting slips, leafy stem cuttings. Both white (*sówkan*) and red (*pɨškano*) varieties are planted.

SOLANACEAE:

Capsicum frutescens. Seeds from this plant are not generally saved, but are burnt along with the fruit. Despite its exclusively ritual use, anybody may ask to be given seed if he is without plants.

Nicotiana tabacum. As soon as the seeds start to blow off the parent plant, they are carefully gathered and stored in a stoppered gourd or discarded glass bottle in the house. No selection of seeds is made; all are sown with great devotion and anticipation.

BIGNONIACEAE AND CUCURBITACEAE:

Crescentia cujete and *Lagenaria siceraria.* Seeds from the previous harvest are saved when the woody outer pericarps of the calabashes and gourds are made into utensils. They are removed and dried in the sun (if the gourd was not left on the vine until completely dry), wrapped, and stored in a small basket. Some Yukpa segregate the seeds according to the various sizes of the fruit of the parent plant, thus selecting for size requirement of the eventual utensil.

Cucurbita moschata. Seeds of particularly sweet and tasty plants are saved after each meal for sowing in the succeeding year. The size of the seed is not considered. The seeds are dried in the sun and then wrapped in a leaf or piece of rag, and stored in a small basket hung from the house rafters.

Sowing, Planting, Care, and Harvesting of the Crops

GRAMINEAE:

Saccharum officinarum (*Sugar Cane*). Sugar cane is planted in clumps, separated by varieties, during March and April. In the past, as many as four varieties of sugar cane have been grown by the Yukpa, but at present only two of these are important: *ķísraš* (white stemmed) and *tínñača* (brown stem with white stripes). Other varieties mentioned as once having been grown are: *stáčapa* and *ķaráķaraš,* but because no specimens were found I am unable to mention their characteristics. Cuttings, about 35 to 40 cm. long, with close nodes and several buds are chosen from younger parent plants. Sugar cane is preferably planted on heavier, water-retentive yet not waterlogged sites, with the cuttings set sufficiently close together so that the leafy crowns will form a closed canopy. In this manner weeds are shaded out, and no further husbandry is necessary. The crop is ratooned and used for three to four years, or until the field is left in fallow.

Zea mays (*Maize*). The Yukpa cultivation cycle revolves around the cultivation of *míšk̦* (*Zea mays*). *Míšk̦* dominates the legends of the origins of the crops; "all of the other crops came with *míšk̦*." Festivities attend the stages of its cultivation, and all of the most important Yukpa celebrations require *míšk̦* (generally in liquid form) as a basic ingredient. In contrast, the recently introduced varieties of maize (*mayíša*) are considered and treated apart from *míšk̦*. *Mayíša* plays no part in either myth or ritual.

TABLE 3

YUPKA CLASSIFICATION OF MAIZE[1]

Yukpa	Race[2]	Color
k̦áyaš wasáčašk̦	*Negrito*	Deep red
k̦omípašk̦	*Costeño*	Yellow/white
k̦orúčašk̦	*Cacao*	Dark blue/black
sarásak̦ k̦anúšk̦	*Común*	Yellow stained with violet
šimíašk̦	*Cacao*	Purple
topómi	*Cariaco* with *Imbricado* crossing	Yellow/white
yik̦ árarašk̦	*Común*	Orange with various colored stains
yuk̦ínk̦ašk̦	*Común*	Rose/violet

[1] This table refers only to *míšk*. The terminology used is that of the Maracá (Yuko) and Irapa (Yukpa) subtribes. The nomenclature used by the other subtribes varies slightly in detail.

[2] After Roberts et al. (1957) and Grant et al. (1963).

The Races of Maize. The 23 races of maize grown in Colombia, together with the 19 distinguished in Venezuela, have been divided according to derivation by Roberts and his associates (Roberts et al., 1957:33) and Grant and his associates (Grant et al., 1963:22) respectively, into three major groups: Primitive, Probably Introduced, and Colombian (or Venezuelan) Hybrids. Six of these races are found in the Yukpa country, classified into varieties according to the predominant colors of the ear. Eight basic types are recognized. No archeological specimens are known from the area.

None of the varieties of maize now grown by the Yukpa appears to belong to Roberts' category of "Primitive Races." According to Cutler,[21] the most primitive variety of maize grown in the area is known as *míšk̦ wahíru* (Guajiro). In 1969, I obtained two small, dark-cherry colored, 12- and 14-rowed ears belonging to what probably is the *Güirua* race (or to a mixture of *Güirua* and *Pira,* or a similar ancient kind of hard flint or popcorn). Among the Yukpa this race of maize is cultivated, to my knowledge, by only one family, the head of which is, biologically, a Guajiro Indian who in his youth was kidnapped and

[21] Letter of November 19, 1970.

raised by members of the Apón subtribe of the Yukpa. A few years ago he moved to Manastará at the headwaters of the Río Negro. Occasionally he returns to visit relatives among the Guajiro tribe, from whom he once obtained this primitive race of maize.

The orange-yellow class known as *topómi* is a mixed group. Basically it belongs to the *"Cariaco"* race of Roberts' "Probably Introduced" category. *Cariaco* is a large-grained, flour maize which is probably derived from lowland varieties grown in the upper Amazon Basin. In Colombia the *Cariaco* race is cultivated in the lowlands of the Caribbean Coastal plain, and the lower Magdalena and Cauca valleys, generally at elevations below 400 m. It is not widely cultivated. By contrast, in Venezuela the *Cariaco* race is widely cultivated at these lower elevations. The Yukpa cultivate it at an elevation of from 800 to 1400 m.

The adaptation of the Yukpa's *Cariaco* variety to cultivation at the higher elevations may have resulted from crossing with maize of *"Imbricado"* race. Some ears of the *topómi* class show a mixture with a pointed, hard-grained, popcorn maize which probably belongs to the *Imbricado* race. According to Roberts and his associates, the *Imbricado* race belongs to the "Probably Introduced" category and may have developed in the highlands of Ecuador, Peru and southern Colombia, where it is cultivated at elevations between 2000 and 2600 m.

Most of the maize cultivated by the Yukpa belongs to the "Colombian Hybrid" category; and in particular to the race known throughout Colombia and Venezuela as *"Común."* This is the Colombian maize best adapted to cultivation in the *t. templada,* and as such is the race most widely cultivated throughout the nation between the elevations of 800 and 1400 m. *Común* is less frequently cultivated in Venezuela, where, according to Grant and his associates (1963:72), it has been much affected by the introgression of improved hybrid strains. The grains are either white or yellow. Because of the lack of careful selection and varietal separation, however, the *Común* cultivated by the Yukpa, by whom it is known as *sarásaḳ ḳanúšḳ, yiḳáraršḳ,* and *yuḳínḳašḳ,* shows some color variation.

The dark-colored *ḳorúčašḳ* and *šimíyašḳ* types belong to the Colombian Hybrid known as *"Cacao,"* which is adapted to cultivation at elevations of from 1300 to 1700 m. This is a slightly dented, floury-grained race which is the type most used by the Yukpa to make *chicha* (*vide infra*).

Komípašḳ is a member of the *"Costeño"* race, one of the races with the widest distribution in Colombia and Venezuela. It is cultivated along the Caribbean coastal lowlands from the Chocó to eastern Venezuela, and in the Cauca and Magdalena valleys from sea-level to an elevation of 2100 m.

Káyas wasáčašḳ belongs to the *"Negrito"* race, which is of very minor importance in Colombia. It is cultivated in the departments of Atlántico, La Guajira, and Magdalena by a few farmers for the production of a bread known locally as *"bollo."* In Venezuela the *Negrito* race is cultivated only sporadically in the coastal lowlands. The Yukpa cultivate dark-red, floury variants of the *Negrito* race with 18 and 20 rows of grains.

Mayíša is a generic term applied by the Yukpa to those types of maize that were acquired in comparatively recent times from the *wátia.* These are all strains of the "Coastal Flint" maize complex which is widely cultivated from the Caribbean littoral of Colombia, along the east coast of South America, to Argentina.

The *mayíša* cultivated by the Yuko was identified by Cutler[22] as a flat seeded white flint of the *Común* race. Similar recent introductions among the Yukpa appear to be derived from the modern hybrid known as "Venezuela no. 1."

A very recently introduced strain of *mayíša* cultivated by the Maracá subtribe is known as *"waípašk̨"* or *"wátia* maize." This is a yellow, hard flint maize with flat kernels and short ears, and is a commercial line which is cultivated on the large, mechanized farms of Costa Rica, Colombia and Venezuela.

Uses of the Races of Maize. The bulk of *míšk̨* grown by the Yukpa belongs to the multi-purpose, floury to soft flint *Común* race. Such maize, together with the *Costeño* and *Cariaco* (*Imbricado*) type is used for routine consumption. Most commonly it is eaten boiled. Less frequently it is eaten roasted or ground up and served warm as flour (*tépese*). Only rarely is this type of *míšk̨* used to make the ceremonial *k̨osíš* and to add a different flavor to *sóña* and *túk̨a.*

The floury *míšk̨* belonging to the *Cacao* and *Negrito* races is used basically for the production of *sóña* and *túk̨a.* When possible, it is used also to make *k̨osíš.*

Mayíša has no ceremonial use. It merely forms part of the routine diet, and is never used for the production of either *chicha* or *k̨osíš.*

Dibbling: Labor. Traditionally the dibbling, and the predibbling preparation of the seed was the occupation of the womenfolk alone. Nowadays this rigid division of labor no longer holds true, and whereas women still aid in its sowing, the situation has almost completely reversed itself, and the task is now virtually exclusively within the province of male labor, as are virtually all non-harvest cultivation activities. There is now only residual female participation in sowing activities. This may be ritually—though it is more likely practically—inspired.[23] Depending on the size of the field and the urgency of the dibbling, however, any combination of competent work group members may sow *míšk̨.*

Dibbling: Procedure and Technology. Planting starts just before the onset of the rainy season, about a month after burning, in March and April. The first crop to be sown is *míšk̨.*

Procedures differ in detail depending upon whether *míšk̨ or mayíša* is being sown. However, nowadays the procedures are standardizing and *míšk̨* is increasingly treated in the less elaborate manner which hitherto was reserved for *mayíša.* Traditionally *míšk̨* is sprouted and then carefully placed in the holes with the radicle pointing downwards; *mayíša* is carried unsprouted to the fields, where it is casually tossed into holes. The *míšk̨* is taken to the field in the basket

---

[22] Op. cit.

[23] During the early stages of fieldwork I thought that the women were forbidden to sow. This belief was actually encouraged by the men, and various reasons for the lack of female participation were put forward. Only when women were seen to be sowing, exposing the lies, and during more profound discussions about ritual and taboo with some of the older men, did the woman's traditional role emerge. Sheepishly the men then admitted that they had been lying. They said that they had switched roles because of the constant ridicule to which they had been subjected by the *wátia,* and especially by the priests at El Tukuku. For reasons associated with masculinity they did not wish to mention this historical role of the women in cultivation. For this reason perhaps few other authors have mentioned this important role of the woman.

in which it was sprouted,[24] which is carried hanging from the shoulder of the arm not wielding the digging stick.

Sowing starts early in the morning. Upon arrival in the field a suitable digging stick is cut and sharpened with a machete. The sower then works lineally up and down the slope making holes (*naškátšk*) and dropping in seeds. Generally the holes are placed to the left and right of the line of movement, so that two almost parallel rows (*tuwísarka*), about 1 m. apart, are planted concurrently. Holes are spaced about 1 to 1.3 m. apart along the rows. The digging stick is punched vigorously into the soil and moved in a circular motion several times to loosen the earth which has been baked by the sun and compacted by torrential showers during the month since burning that it has lain bare. Sowing the corn and periodically halting to resharpen the digging stick, the sower proceeds slowly, carefully pulling up (*yuwičɨr*) the many weed seedlings which sprout and attain the height of 15 to 25 cm. between planting and burning. Into the 7 cm. deep hole are placed from four to six seeds. This many seeds, which are never thinned, are used as an insurance against losses to "worms." The holes are left open because the Indians believe that the seeds would die if they were covered over. According to Santelos' observations (1959:269) losses of the seed to birds and animals are therefore enormous, especially because neither watchers nor scarecrows are used to keep the marauders away. In the area occupied by the Maracá seed loss to birds (*tímsaš nimápinini*) does occur, but it is never as serious as the loss to "worms" and seed-carrying ants. The Yukpa themselves hardly ever blame birds for the failure of plants to appear after sowing.

The sowing proceeds until the whole of the first year field has been filled, or until the supply of seed is exhausted. If there is plenty of seed remaining after the new field has been completed, the cultivator may decide to plant some in a second year field, even though he knows that the yield will be about half that of a first year site.

In an effort to prevent cross-pollination, the different varieties are sown at intervals two weeks apart, so that one is several centimeters high before another is sown. Hence the different varieties may be planted in adjacent rows, because temporal rather than spatial separation is used by the Yukpa. *Mayíša* which occurs only in white and yellow varieties, is planted apart from the *míšk*, but within the same field. This segregation of the different varieties into distinct blocks of rows is also useful because the harvesters have little difficulty in locating the desired type of corn when it is needed. Further, staggered planting means that the harvest period is extended, and that fresh corn is available over a longer time than would be the case if all the corn was planted at once.

The main reason given by the Indians for the varietal segregation is that the hybrids resulting from uncontrolled crossing give progressively smaller ears. Therefore, over the years there is a marked reduction in yields. Despite their intentions, however, the maize grown by the Yukpa is quite mixed, and indicates that their efforts at varietal segregation are of little avail.

Post-Dibbling Taboos. After the completion of sowing until the *míšk* has grown to a height of about 60 cm., the women sowers must observe *višípra*, a taboo

[24] Today any suitable container, such as an old paint can discarded by the *wátia*, enamel household bowl, or aluminum cooking pot may be hung from a string or carried in the hand and used to carry the seeds. I have never seen a calabash or gourd used for this purpose.

which forbids them to cook, and *vorípaš tomásay,* a taboo against touching "strong smelling plants, and plants which smell like onions." Observance of these taboos is said to prevent the "worms" from endangering the crop. Men are obliged to observe a similar taboo against handling certain plants after they have completed the sowing of beans.

The Matríčay Celebration. A small celebration is held when tiny, immature ears (*matríčay*)[25] have formed on the *míšķ*. Some ears, together with the meat of a small mammal or bird, are ground up and made into a special type of *ķosíš,* known as *sapóčpo.* The ground *míšķ* and meat is replaced into the husks, conserving carefully the original shape of the ear, and all is tied with a cotton thread. When cooked, the binding thread, which is also known as *sapóčpo,* is carefully removed and saved. The *sapóčpo* are chewed first by the women who sowed the *míšķ,* and spat onto the inside of each elbow. This is done to insure that the arms will be strong enough to permit successful sowing again during the following year. After this brief ceremony, all may partake of the food. The thread which was used to secure the *sapóčpo* is then carefully tied around the wrists of those who will make the fields for the following season's *míšķ* crop. The *sapóčpo* is worn until the start of field clearance. It is worn to ensure that the cutting will go without a hitch, and that the *míšķ* subsequently sown will yield well. Perhaps, too, it symbolizes the continuity of Yukpa life, centered upon the repetition of the *míšķ* cycle.

Weeding. Traditionally corn in first year fields was never weeded, regardless of the intensity of weed infestation.[26] When the field was cut in high forest, the Yukpa claim that no weeding was necessary during the first year of cultivation. They admit, however, that a field cut in low forest fallow was done only occasionally under the traditional system. Second year fields are weeded.

Thinning. The young maize is never thinned, and as a result, four or more spindly plants often grow from the same hole. However, this too is changing as the Indians have a greater cognizance of improved methods, and some now sow fewer seeds per hole.

Filling In. This is done only rarely in the few places in which the maize fails to sprout. Some of the surplus seed may be saved by the more foresighted members of the community for this purpose, but generally any seed left over from the initial planting has long since been used to make *sóña* or *ķosíš.* Generally root crops are planted in the gaps left in the corn rows (as distinct from interplanting).

Practical Care of the Growing Crop. Apart from the gathering of corn "worms," and hunting to reduce the predator population, the growing corn is not cared for by cultivation techniques. Occasionally the state of the corn is checked by the womenfolk, who, during the scarce season, carefully comb the field for edible insects and items of the natural vegetation. The young corn is a good source of a variety of corn "worms," which are carefully collected, wrapped in

---

[25] The term *"matríčay"* is also applied to play arrows. These consist of a shaft of cane tipped with a kernel-less corn cob.

[26] Under the influence of *wátia* techniques this is now changing, especially around the missions. Weeds are cut with a machete, and piled to dry at one side of the field. They are then burned at the beginning of the second year of field use, together with the newly cut slash.

leaves, roasted and eaten. In this manner some of the worm damage to the corn may be reduced. However, among the invertebrates gathered by the Yukpa, "worms" are really only an emergency food, which, under the influence of *wátia* ways, is of declining importance. Periodically men and women examine the young ears for infestation. A tube is made from leaves of the *ḳamása* (*Carica microcarpa?*), and filled with "worms" and then it is tossed far away.

During the growing season many of the corn leaves become yellow and withered (*ipčane*).[27] The Yukpa believe that this is a result of overcrowding, and furthermore that it is contagious. Such leaves are carefully stripped and thrown away so that the other plants will not be infected.

The cultivated field is a relatively rich hunting ground (Ruddle, 1970*b*). Early in the growing season the men of the work group construct hunting platforms (*šiḳárto*) in the field. Other than this, no form of watching, scaring, or guarding in the growing corn is undertaken.

Ritual Care of the Growing Crop. One of the rituals involved in caring for crops is rainmaking. At the beginning of the growing season, when it is particularly important that the *míšḳ* receive adequate moisture, and the onset of the rainy season may be delayed, the Yukpa repair to a number of sacred pools, into which stones are pitched.[28] While pitching the stones they ask the rain spirit, *imánta,* to send the gentle, cold rains, which arrive in March and April, before the onset of the main rainy season. At any time during the growing season, when there is danger of the *míšḳ* drying out, the same ritual is repeated.

A second ritual is performed to drive out *iḳanmɨráḳtay.* At any time when the *míšḳ* seems to be doing poorly, the head of the household goes to the field carrying a small gourd full of dried red peppers. He builds a small fire in the field, and carefully burns them. The malevolent spirit, *iḳanmɨráḳtay,* who lives in the soil and is endangering the *míšḳ,* detests the smell of red pepper smoke and quickly flees the area, leaving the crop to grow in peace.

The Harvest Festival. The harvest time for *míšḳ* is the high point of the Yukpa calendar. It is at this time that community solidarity is reemphasized. The locus of each celebration is a work group chief's household, but the focus is very definitely on the community.

The extended harvest festival is really a tripartite affair, consisting of the following distinct phases. of which the first two are regarded essentially as preliminaries: first, *ḳosípo osimáytpo,* at which *ḳosíš* is the principal item consumed, liberally washed down with *sóña,* prepared from the *míšḳ* remaining from the previous harvest; second, *túḳapo osimáytpo,* at which *túḳa* alone is prepared; and third, *soñápo osimayḳítpo,* which is devoted entirely to the imbibing of *sóña.*

Each part of the festival is separated by a period of from 20 to 25 days (or perhaps even longer) depending on the number of work groups within a subtribe. The heads of each group are expected to give all three celebrations (*osimáytpo*), in the chronological order indicated above, at their houses. The

[27] The cause of the phenomenon was not discernible.

[28] For example, Okorátmɨ at Socorpa, and the Lake of Kunaná near the village of the same name in the upper Negro Valley.

host groups follow a predetermined order for all three parts of the harvest festival.

Each group's celebration continues day and night for two or three days, depending on the quantity of refreshments provided, with a lapse of three to four days between parties, during which time the participants recover from severe hangovers, bandage their wounds, bury their dead (shot with arrows and poisoned), prepare their weapons, and generally organize for the next party.

Kosípo osimáytpo: Before the actual preparations start for each party the males of the host group go hunting, to provide sufficient meat for the preparation of *ḳosíš*, and for ritual reasons. All of the meat is used in the production of *ḳosíš*, but initially some is reserved for the *míšḳ* to "eat." The maize's meat is placed in the bottom of each harvest basket, and the group women go to the field to collect mature but still green *míšḳ* (*wɨrápa*). The maize is picked and placed on top of the meat in the basket. The return journey from the field to the house gives it sufficient time to eat its fill of the meat in the basket. This gift of meat is thought to please the *míšḳ*, so that it will willingly produce bountifully in the year to come. After fulfilling its ritual role, the meat in the basket is added to the rest of the meat which is then ground up, added to the ground *míšḳ*, and made into *ḳosíš*.

At this time *sóña* is prepared and used for liquid refreshment during the consumption of the *ḳosíš*. Although all the Indians still manage to become thoroughly drunk, the primary objective of the celebration is the consumption of *ḳosíš*. The *sóña* orgy comes later in the party sequence (*vide infra*, pp. 91–92).

Kosípo osimáytpo plays an integral and important part in the harvest festival. It must be emphasized, however, that its role is definitely only tertiary to the other phases, and it is with the celebration of *soñápo osimayḳítpo* that the festival reaches its climax.

Túḳapo osimáytpo: Túḳa is traditionally prepared for consumption only once a year,[29] during the second phase of the harvest festival.

Preceding the collection of *míšḳ* for the preparation of the *túḳa* the male members of the work group have to go hunting for meat to be used, exclusively this time, for the maize's ritual meal. Very early in the morning of the day appointed for the start of the celebration large baskets are prepared, and the meat for the *míšḳ* is placed in them. Then the men dance the *menúri súwaḳ* with the baskets on their backs. The work group women stand singing to one side of the house clearing. They do not join the dance. Conversely, the men do not sing, instead they shout three times during the dance. After the *menúri súwaḳ* the women take the baskets and proceed to the field to collect maize. Later the men follow the women to the field, where they relieve them of the full baskets, which are too heavy for them to carry. The baskets are then taken to a site alongside the field where they are emptied of their contents. Before the shelling commences the baskets are thrown a short distance away from the gathering *míšḳ*. This is done because it is said that their close proximity to the shelling operation would cause the picked *míšḳ* to turn bad (*árḳna sayáta menúri*), and thus more would have to be collected for the celebration. When the *míšḳ* is

[29] This is now changing, at least in the vicinity of the Catholic missions, because the priests have forbidden the production of the heady *sóña* and have encouraged the substitution of *túḳa*. *Túḳa* therefore is now produced throughout the year.

completely shelled, the baskets are retrieved and filled. The men then carry them back to the house shouting all of the time. Once back at the house dancing starts again. The men alone dance and shout, carrying the full baskets. Again the women are limited to singing an accompaniment.

After the dancing, the full bags are first taken to any house other than the host's, then, singly, they are returned to his house, where the *túḳa* preparation is to take place. It is said that if all the bags were taken together, and directly from the dance to the host's house, then the grains would not yield enough *túḳa* for all the participants.

While the women grind the *míšḳ* and place it in boiling water the men start to dance and shout again. When boiling the *míšḳ* the women have to carefully observe *oyúr pumána,* a taboo against letting any of the contents of the pot burn. If they burn there is thought to be a strong possibility that the host would become involved in an argument during the course of the festivities and a killing would result. While the mixture is heating it is carefully stirred with a stem of millet. Millet is used because it is thought that some of its strength will pass to the stalks of the maize still remaining in the field (*piríḳo san sintróḳaḳ*), and so the remainder of the harvest festival will not be jeopardized through the loss of maize to wind damage.

As the preparation progresses the brew is poured into a number of very large gourds. When these are full, the contents are emptied into the *chicha* trough (*ḳanówa*). Sometimes mashed sweet potatoes or bananas are added to the contents of the trough to modify the flavor.

The first evening of the celebration is devoted entirely to dancing and singing. There is no drinking, as the *túḳa* is not yet ready. Further, although there is great temptation to taste the brew, the taboo against doing so (*yipɨ́ni antási*) is enforced, for to taste the brew prematurely would cause heavy rains to fall during the night, and that would ruin the dancing.

Early in the morning of the second day the *túḳa* is ready. The host is the first to sample it. He puts a piece of meat in his mouth, and then swallows it together with a gulp of the *túḳa*. Meat and *túḳa* are then given to all of the guests. In the meantime the women start to prepare another batch of *túḳa* for the second day of the festivities. Drinking then commences in earnest, and by midday the whole company is quite mellow. "*Túḳa* is sweet, and it doesn't make you rotten drunk like *sóña* does, rather it makes you feel nice and lazy." Before sloth takes too great a hold of his guests, the host insists that they bestir themselves and start to dance again. "Dancing is very important, and you have to do a lot of it, because it makes the *míšḳ* content and so next year's crop will be bountiful." In the early afternoon everybody is too lethargic to continue dancing and, besides, it is too hot. Everybody goes to sleep. Silence reigns.

Drinking and dancing continue for three days. After the third and final morning's *túḳa* has been consumed the men repair to the host's corn field to seek *míšḳ* for a meal. The host is forbidden to pick the *míšḳ,* so he directs the others, and then supervises the sharing. The others offer him ears for a meal. The *míšḳ* secured, the men return to the host's house, where once again they start to dance, while the womenfolk cook the ears. The eating of the *míšḳ* concludes *túḳapo osimáytpo*.

Túka traditionally was prepared for human consumption only once yearly, for use in the second phase of the harvest festival. In contrast, *sóña* was prepared for every other celebration, the most important of which was *soñápo osimaykítpo*.

Soñápo osimaykítpo: This represents the eagerly awaited climax of the harvest festival and the first of many occasions on which *sóña* is prepared with the new *míšk*. The preparation of the *sóña* starts nine days before it is required for drinking.[30] Again the men must hunt in order to provide the meat for the maize's ritual meal, and the women, with the meat in their collection baskets harvest *míšk*.

The first day of preparation is devoted entirely to grinding the *míšk* to flour (*míšk tépese*). After all has been ground the women then gather the leaves of a *štoíyi* (*Ficus urbaniana?*) tree. The flour is wrapped in these leaves and bound with any available, nontoxic vine (*mimíšk*). Balloonvine was formerly widely used for this purpose. The resultant small packets (*tamúyas*) are then boiled in water for one day. While the *tamúyas* are boiling, a rack (*tamúyas yipónš*) of sticks is made inside the house. The *tamúyas* are removed from the boiling water, unwrapped, and set upon the *tamúyas yipónš* to dry for four to five days. When they are thoroughly dry they are rewrapped and placed in the trough with cold water to soak for one day. After they have been softened by this soaking they are removed and ground once again. This second grinding requires a further day's labor by the women. The reground flour is then replaced in the trough, which is filled with water and stirred thoroughly. At this point sugar cane juice is added to the mixture. The top of the trough is then covered loosely with the leaves of plantains and bananas. The Yukpa never masticate the maize to start the fermentation process.

The women then go once again to the corn field where a couple more baskets are filled with *míšk*. From this second batch of *míšk* more *tamúyas* are made. These are placed directly into the trough, "to sweeten the brew." At this stage the contents are sampled to see how they are progressing. More leaves of plantains and bananas are cut, and the mixture is stirred thoroughly for a second time. Then a top (*yɨtáprɨr*) of leaves, tightly secured with vines, is placed over the trough. The stated purpose of the snugly fitting top is to keep the "warm smell" inside because, "if it gets out the *sóña* will be weak and cold tasting. *Sóña* has to taste hot." In reality, of course, the tight top creates the necessary anaerobic environment for the hydrolysis process to take place.

Early on the first morning of *soñápo osimaykítpo*, before the guests start to arrive, the *witimímo* are hung over the trough. They are prepared for use in the harvest festival before the *míšk* is ready for harvesting, and are kept in the host's house until the start of the celebration. They consist of seven small, rectangular, wooden tablets, a larger one being called *yišámpi*, or mother, and six smaller ones, *sinpáne*, which are arranged three on either side of the *yišámpi*. These are tied on cotton threads, and suspended from a horizontal pole (*viš*) so that they hang over the trough, and swing gently in the breeze. Designs, which represent worms, are drawn on the tablets and on the poles with charcoal, to enable them to "see" the spirits of the dead Yukpa (*okátu*) lurking in

[30] De Booy (1918*c*:202) gave a different method and length of time for the preparation of *chicha*. There is considerable confusion surrounding this subject in the literature.

the bushes around the homestead. It is said that the tablets keep the spirits away from the trough. If *okátu* get into the trough there is grave danger that they will be swallowed by the drinkers. This is serious, as *okátu* endangers the stomach, and causes severe attacks of vomiting, from which one can die. The tablets are always present at the celebration so that all members of the community may partake of the *sóña* without fear. On other occasions when *sóña* is being drunk tablets need not be present. When they are absent, however, women and children may not participate, as they are the ones most endangered by the presence of *okátu*. When the *sóña* has been completely consumed the tablets are removed and hung from the lintel of the host's door. The poles are thrown away. The designs are scraped off the tablets, which may be used one further time only.

When the guests are all assembled they paint their faces red with the juice of the seeds of annatto mascerated in water. Formerly the sap of another plant, *kútia* (*Amaranthus cruentus*) was also used. It was sown near the houses, and was ready for cosmetic use when the *míšk* was mature.

When all is ready, the host cuts a rectangular opening (*tomáčkas*) in the *yɨtáprɨr*, and folds back the lid (*túksi*), thus made. He then serves *sóña* to the assembled company, using a server made from half of a gourd or calabash.[31]

The community then proceeds to drink, dance, sing, and later to sexually assault the womenfolk, fight, wound, kill each other, and to generally settle old scores, until the *sóña* has been entirely consumed, whence they collapse into a drunken stupor (Reichel-Dolmatoff, 1945:66; Ruddle, 1970*a*:129; Schön and Jam, 1953*b*:71; Gusinde, 1956:203).

So ends the harvest festival and the ritual activities for the annual cultivation cycle. Harvesting continues, but now the community's main thoughts are turning to the preparations for the following season.

Activities during Maize Ripening. After the harvest festival the remainder of the maize is left standing in the field for several weeks to dry thoroughly. Women go daily to fields close to the house, or less frequently to those more distant, to pick enough ears for immediate use. Only the women pick corn. The taller plants are cut at the base with a machete and the cob is pulled off. The husks are stripped in the field (if *kosíš* are not desired). The ear is pulled off and the stalks are left standing. Stalks which are still green may later be harvested by the men and fed to donkeys and mules.

Progressively the amount of maize remaining in the field is reduced. During this time the loss and damage of the drying grain, particularly by corn borers, monkeys, and rodents is substantial. However, the Yukpa are not upset by this as the maize field at this season becomes an excellent hunting ground. They regard the grain lost as bait, and feel that they are amply compensated by the greatly increased meat supply that becomes available. This period is also the time for the repair and construction of the storage house (*oyíta mosmɨna*), corn cribs (*karáčašk* and *tipértartos*), and harvesting baskets. These jobs are undertaken by the menfolk.

Ritual Aids for Drying Míšk. At this time of the year when the maize is drying in the field there is great concern about unduly heavy rainfalls. Although the

[31] In the Yukpa communities of the *t. caliente*, the ligneous spathe of the *kurúkmatoro* (*Scheelea* sp.) palm is used as a *chicha* server. This utensil is also called a *kurúkmatoro*.

season of heaviest rains is passed, the crop may be substantially reduced by rotting. It is during this time of the year that ritual is employed to drive away the clouds and to make the sun shine to dry the *míšk̨*. Two methods are used to chase away the clouds. Parts of plants, especially the fruit of *k̨ónt léla* (*Hibiscus abelmoschus*), *mapíca* (beeswax mixed with the ashes of gourd [*Lagenaria sicareria*]), the seeds of gourd, the leaves of *k̨amúr biyáiya* (*Psychotria* sp.), *ok̨átia* (*Wedelia* sp.), which also serve to drive away spirits in general, together with the leaves of ferns (*Diplazium* sp., *Dryopteris* sp., *Polypodium* sp., *Schizea* sp.) are burned on the domestic fires. The smoke of these plants is said to be very effective for dispersing clouds rapidly. If no suitable plants are available, the clouds may be blown and pushed away, although this is not considered to be as effective. To this end elders in the communities may be seen cupping their hands to their mouths and gently blowing and pushing their arms skyward on rainy days in August and September.[32]

Harvesting. When the men judge that the maize remaining in the field is sufficiently dry, and that it had better be harvested before too much more is consumed by predators, the women are sent to bring it in. It is brought to the house or storage shed little by little each day by the women. At the house the men undertake most of the work, selecting first the seed corn to be dried and stored apart from the rest of the harvest. After the seed has been selected the remainder of the maize is shelled (*tínk̨as*) and spread out on a mat in the sun for further drying (*apóte intášpoš*). At the day's end the dried grain is then poured into the crib. Sometimes the grain is stored on the ear. This depends on the preference of the work group head, and also reflects a liking for corn on the cob to supplement meals of *k̨osíš* and *tépese*. Thus the daily round of picking, transporting, shelling, drying, and storage occupies the work group until enough maize has been stored.

Storage. If the dwelling house is a considerable distance from the corn field, the work group constructs a storage house in or near the field. This saves the heavy labor of transporting the entire harvest within a concentrated period to the dwelling site. It is much easier for the women to go to the store now and again during the year to collect enough for the meal, especially as they have frequently to go to the field anyway to harvest the root crops. The separate store house and cribs are gradually being rendered obsolete by the increasing use of donkeys and mules for transportation, and sacks and lofts (*arápa*) within the dwelling house for storage.

Within either the dwelling house or the store house corn cribs are constructed. These are of two types, one of a simple wickerwork frame to contain corn ears and the other enclosed with broad leaves to hold grain (pls. 6 and 7). These cribs are totally ineffective in preventing the loss of corn to animals and insects. Despite the fact that deadfall rattraps (*manú*) are set beneath the cribs the loss and spoilage of stored grain is considerable. However, it was frequently repeated that losses of grain stored in cribs in the dwelling house were more than those from the store house, because of the activities of domestic animals and fowl, who helped themselves liberally to the readily accessible supply of food.

---

[32] Bolinder (1958:147) also noted this.

To guard against the theft of stored corn by other Indians, two anthropomorphic posts, *yimópe* (father),[33] and *tómayra* (priest-shaman),[34] traditionally were stationed on either side of the cribs (S.C.N.L.S.: 1953:pl. 31). These were carefully scraped to a lightish color with machetes so that they would be able to "see" and scare away potential thieves. If these two posts were not erected passers-by would have no compunctions about stealing stored corn.

Losses to domestic animals have been reduced by storage of sacked corn in lofts in the dwelling, but this has led to an increase of the rat population around the houses. With typical aplomb the older Indians have turned this situation to their advantage, and the practice of fireside hunting is now widespread. Interestingly, only a few domestic cats (*Felis catus*) are to be found in Yukpa communities. Traditionally the Indians have preferred to use rats as a source of human rather than pet food, although this predilection is now rapidly disappearing under the influence of the *wátia*.

Handling of Unusually Abundant Crops and Surplus. The women continue to bring the dried corn from the field until either all is harvested or a sufficient amount to keep the work group until the following harvest has been stored. Then they tire of the work. In the past surplus corn was left standing in the field to rot, and to entice animals for the benefit of hunters. After a work group had stored sufficient grain, other groups whose harvests for various reasons may have been poor, were invited to help themselves to the surplus. They were not permitted to harvest what remained in the field in one fell swoop, but rather they came daily, or as hunger dictated, to fill sufficient baskets for immediate needs. No payment of any description was required for this privilege.

Today the situation has changed somewhat. Whereas the fortunate are still obliged to help the poor, a market mechanism is developing, and the Yukpa can now sell their small surpluses to the nearby missions, towns, *haciendas,* and *fincas.* It must be emphasized, however, that in the normal course of events scarcely any of the work groups provide sufficient for their own needs, and there is a regular season of scarcity. Only a very tiny minority of exceptionally industrious and acculturated individuals produce an occasional saleable surplus.

Oryza sativa (*Rice*). Upland rice cultivation was unsuccessfully tried by one Maracá during 1969. The seed was given to the cultivator by the Capuchin priest, but because nobody knew how the crop should be harvested, the harvest was lost.[35] The first experiment with rice cultivation had the same result as other attempts to introduce new crops to the Maracá. The priest at Socorpa has at various times tried to induce the Indians to cultivate cabbage, peas, onions, and many other similar crops. All attempts failed, however, because the Indians did not appreciate the need for these additions to their inventory of cultivated plants, and therefore did not care to bother with the extra work involved in good hubandry of crops which added nothing special to their food supply. These crops do well at Socorpa in the mission kitchen garden where they are weeded, thinned, and cared for by Maracás working under the close supervision of nuns.

[33] An authoritarian figure in Yukpa society.

[34] *Vide supra,* p. 33.

[35] Personal communication, R. P. Hermenegildo Pérez, December 1969.

Conversely, as a result of working in the mission maize fields, the Maracá now appreciate the usefulness of controlled burning, thinning, and weeding. Such lessons are rapidly learned and, within the limits of work group capability, are applied with increasing frequency and effectiveness. This is not surprising because improvements in maize cultivation touch on the crops long basic to Yukpa life. The new techniques are readily perceived as beneficial and are easily integrated without undue disruption. This may not be the case with attempts to introduce totally unfamiliar crops.

Panicum miliaceum (*Common Millet*). Stems of this grass are used principally for making arrow shafts and the seeds used as bird bait (*wopɨ́k̜š*). Millet is planted in clumps alongside trails in fields and around the house site (pl. 8*b*). Plantings close to the house have the advantage that birds, attracted by the seeds of the mature plant, can conveniently be shot from the comfort of the house.

The seeds are dibbled in March and April, coincident with *míšk̜* sowing, in many small clumps, the proximity of the plants depending on the desired thickness of the stems. When their stems are required for making lightweight arrows used for shooting birds at a short distance, the seeds are dibbled close together. Conversely, when thicker stems are needed for heavier arrows (but not as heavy as those made from wild cane), the seeds are spaced farther apart (Ruddle, 1970*b:* 26–32).

Seeds are harvested in September and October. Some are selected for the following year, but the majority are put aside for use as bird bait. Alternatively some cultivators like to leave a few plants *in situ,* so that they may have a site in the field for hunting birds; and also in order that the millet may propagate themselves for the following season.

Cutting of the stems can take place at any time during the growth period, that is, when they have attained the desired dimension for arrow shafts. The major harvest takes place in August, September, and October. Only selected stems are cut, the criteria being straightness and strength.

Gynerium sagittatum (*Wild Cane*). This is sown only by cultivators who live at the high altitudes, where it is not encountered in the wild state. Those who live in the lower Tukuku and Negro valleys, for example, have no need to sow it because it occurs in abundance in humid riverine sites.

The Maracá collect some cane seeds when on a visit to the lower lying areas of the Río Maracá and Río Roncón valleys. The seeds are sown immediately upon their return in areas of fallow alongside the main river. Cane is sown only by women, for it is said that when cane is sown by men too many flowers result and the stems are unsuitable for arrow shafts. The cane is sown at any time during the year and, similarly, harvested whenever the need arises for heavy arrow shafts. Some of the Maracás prefer not to have a planting near the community, as they would rather have an excuse to make three- or four-day cane collecting expeditions to the lowlands.

Hyparrhenia rufa (*Jaraguá*), *Melinis minutiflora* (*Molasses Grass*) *and Panicum maximum* (*Guinea Grass*). These three pasture grasses are sown on a modest scale by those few Yukpa who are gradually developing an interest in cattle raising. There are also tiny isolated patches of "artificial" pastures sown by broadcasting at the end of the cultivation sequence, to provide forage for their burros and mules. Seed, initially obtained from the *wátia,* and in particular from the Capuchin missionaries, is nowadays generally gathered from pastures sown at an earlier date.

ARACEAE:

Xanthosoma sagittifolium (*Yautia*). Yautia (*sɨríašk*) is interplanted between rows of maize, plantains and bananas, and, as with all crops, is also grown in monocultural stands (pl. 8*a*). It is spaced about one m. apart. Spacing is always very approximate and depends largely on the size and lie of unburned trunks, stumps and rocks.

For planting the soil is loosened with a *coa,* and corms, cormels, and cuttings made from the top of the corm together with the lower 20–25 cm. of the petiole, are planted to a depth of about 30 cm.

The Yukpa feel that there is no great urgency to plant the root crops, which may be successfully planted at any time after the height of the dry season. There is no specific time for the planting activities, with the exception of the transplanting of taro. After the completion of taro transplanting the sowing of the grain crop assumes first priority; however, during the sowing a certain amount of interplanting of root crops takes place. There is, for any given work group, an overlap of sowing and planting activities, especially during the interval between the sowing of different varieties of *míšk,* but the bulk of planting takes place after the *míšk,* sowing is completed, and before the heavy rains of May.

The Yukpa distinguish between "sweet" and "hot" varieties of yautia which are planted separately for convenience during harvesting. Six "sweet" varieties are recognized: The first is *tipípus.* This is grown mainly for its leaves which are considered to be especially tasty. This variety is sown on sandy soils which have good drainage. The second variety, *waátu* also grows best on sandy soils. It is distinguished by having a "pinkish" corm. Third is *aatíya.* The corm of this variety has a yellow coloration. It too grows best on well-drained soils. The fourth variety is *mačuára.* The corms are white to cream colored and yield best in sandy soils. The fifth variety, *soátus,* is distinguished by its white corm, which needs both full sun and rich black earth. Unlike the first four varieties it does not yield well when interplanted with manioc, plantains and bananas, nor when planted in sandy soils.

Only one "hot" variety of yautia, called *šíričayi,* is planted by the Yukpa. It requires deep, rich, black soil and full sun to yield well.

Yautia which is interplanted in first-year fields is not weeded until harvesting starts. As each plant is harvested, the surrounding area is thoroughly cleaned before the new corms and cormels are inserted. The weeds are either pulled out with the hand or cut with a machete, and the soil superficially dug over with the point of the machete to remove the roots. The weeds are left on the soil surface as mulch.

Harvesting is done by women as often as necessary to satisfy immediate food needs. If the yautia is growing at a distance from the dwelling, enough may be harvested for two weeks' meal requirements. In that case it is stored in a hole dug in the soil (*nóno tɨnkas*) close to the house.

Colocasia antiquorum (*Taro*). Taro (*ḵɨn*) is planted by the Yukpa in the Tukuku and Negro valleys, but not by the Maracá. Planting is handled in two ways. Some cultivators make a wet pit and transplant the young plants from it to the field, whereas others plant the clones directly, in the same field as manioc. The wet pit is used only twice in one year and then abandoned.

Planting, weeding, and harvesting procedures are almost identical to those for yautia. Although the variety of taro planted by the Yukpa does best in moist soils, it is also tolerant of sandy, stony soils and shady sites.

BROMELIACEAE:

Ananus comosus (*Pineapple*). This is another minor crop of the Yukpa which was introduced comparatively recently to the tribe. It is usually grown exclusively around the house site, although occasionally it is extended into a field. The plants thrive on well-drained and sandy house sites, because they are intolerant of waterlogging. Production is generally poor, however, because the plants are spaced too far apart to both physically support each other and to shade out weeds.

AGAVACEAE:

Agave viridis (*Agave*). The leaves of the agave (*sapára*), which, together with *Furcraea gigantea,* is common in the savanna areas, provide fibres which the Yukpa use to manufacture string and rope. The leaves are also occasionally used for makeshift roofing. For the convenience of the women, whose task it is to make the string and rope, young agaves are sought in the savanna and *rastrojo,* and planted close to the dwelling, where they are treated as part of the cultivated vegetation.

No criteria are used in the selection of the plant. The plant has merely to be small enough to be easily transported for replanting. Agaves used in the male birth celebration (*mášuḵ ápa wátpo*) are carefully selected, for the size and number of the leaves, in the savanna (Ruddle, 1970*a*:269). Those replanted near the dwelling are never used in this ritual.

DIOSCOREACEAE:

Dioscorea alata (*Yams*). Yams are only of minor importance in the cultivation system. Plants are scattered haphazardly both in fields and around the house site. Stakes are not provided for the plant to climb, but rather they grasp other crops and weeds for support. Occasionally a yam plant is trained to climb a convenient tree stump (pl. 9).

MUSACEAE:

Musa spp. (*Plantains and Bananas*). Plantains and bananas are planted in March and April. Suckers are cut from the base of mature plants and any roots which may remain are trimmed off with a machete. As a general rule they are planted in damp and sheltered spots where rich alluvial, black earth is available. Fairly level areas along small stream courses are considered to be ideal sites; for the surrounding steeper slopes and trees protect the plants from wind damage, and the soils are sufficiently rich and humid. After the best stream course sites have been planted, other, less suitable, sites on steeper slopes and raised stream terraces are utilized. Such sites represent a compromise, for the richer rendzina soils are generally encountered in localities prone to wind damage. Another compromise solution is to plant bananas and plantains on the *sásare* soils of the raised terraces, but the plants are somewhat retarded by the poorer soils, and further, the terraces are prone to undercutting and collapse in the rainy season (pl. 1).

Plantains and bananas are planted both scattered around the dwelling and in the fields. In the fields they are not grouped according to variety, because the varieties are easily distinguishable.

Four varieties of Musaceae, which are generally baked, but sometimes boiled, before eating are:

k̦uràntanašk̦	(*Dominicos*)[36]
misínasa	(*Banano*)
pirípria	(*Santa Maria*)
šúk̦iapa	(*Bocadillo*)

All of the other varieties are eaten both uncooked as deserts and cooked. They comprise:

čapɨčápk̦ano	(*Felo*)
k̦ak̦ɨ́k̦aši	(?)
k̦amčantáne	(*Guineo Sentado*)
k̦iníya	(*Guineo de Monte*)
metetpa	(*Manzana*)
wasántotč	(?)

Holes are made with a *coa* in second year fields and the plants spaced about 2.5 m. apart in rows. They are weeded once a year only, except when intercropping is to take place. The weeds are slashed with a machete and left where they fall to mulch. Burning is not done for fear of damaging the growing plants.

Harvesting of the first bunches takes place nine to 13 months after planting, depending on the locality. The plants will continue to yield for about 15 years, and so very old stands are to be found around the larger established house sites. Smaller quantities of plants set out in fields are generally abandoned after three or four years, when the field is fallowed, whereas larger areas, planted in monocultural stands, will be maintained for many years. Increasingly, monocultural stands of older plants are being interplanted with coffee. However, it is the commoner practice among the acculturated Yukpa to clear tall forest fallow, leaving an adequate number of shade trees for coffee plantations.

CANNACEAE:

Canna edulis (*Canna*). Historically canna was widely cultivated in the highlands of the New World. Among the Yukpa it was an important root crop, its young tubers providing a welcome supplementary source of starch. Although the boiled tubers have a pleasant sweet taste, they are somewhat fibrous, and so are considered inferior to sweet potatoes. The crop is now of little importance among the Yukpa as a food crop and few plants are cultivated for this purpose.

[36] The local Colombian names are given in parentheses.

Recently, it has assumed the role of an ornamental hedge plant (pl. 11). In this the Yukpa have been influenced by the *wátia* who commonly use the plant for this purpose. Formerly the plants were planted and treated in the same manner as other root crops. Harvesting occurred at intervals during the year. Currently harvesting is practiced only in case of dire emergency.

AMARANTHACEAE:

Amaranthus cruentus (*Amaranth*). This annual herb was once widely cultivated by the Yukpa for the production of a red dye used as a ceremonial facial paint.[37] The Indians disclaim any knowledge of the use of seeds or leaves for food in earlier times.

The plant is still commonly encountered around the dwelling sites, in the *rastrojo* and in the fields where it is both cultivated and thrives as an escape. Although now little used for cosmetic purposes, the amaranth is still sown by the more conservative older folk. For sowing the seeds are simply broadcast casually across the area to be planted.

LEGUMINOSEAE:

Phaseolus vulgaris, P. lunatus and Dolichos lablab (*Beans*). Climbing beans have long had a place in the Yukpa inventory of crops. Bush beans, it appears, have all entered from contacts with the *wátia* over the last half century. The beans most widely cultivated among the Yukpa are *P. vulgaris* and hybrids of this bean with introduced types. *P. lunatus* and *Dolichos lablab* are not widely cultivated.

Four different methods of bean cultivation are practiced: Under the first, climbing beans are planted alongside maize plants after they have attained a height of about 7.5 cm. and the bean is later supported by the maize stalk; under the second, special trellises (*tapótritos*) are constructed close to the houses (pl. 10) on which beans are trained; under the third, bush beans are frequently intercropped with maize; and under the fourth, the most commonly used method, a separate field is cut for bush beans. Sometimes the bean field is prepared at the same time as general field clearance is taking place, but, copying the *wátia*, some Yukpa now prepare a separate bean field later.

Beans are quite tolerant of soil conditions and are even grown occasionally on the *vípopa*. They will not, however, tolerate waterlogging. The fields are not weeded.

When mature, some beans are brought to the house for immediate food needs, but the bulk of the crop remains in the field to dry. The chief concern is to have beans shelled and stored before they are dampened by rain. Generally therefore, beans are not thoroughly dry when they are brought in from the field. Unshelled beans are spread out in the sun on a sleeping mat to pop open. To aid in this they are beaten with a stick. Beans are also shelled by hand (*ɨtátķas*). Men, women and children all share this job. The shelled beans are then once again spread out in the sun to ensure complete drying. Dried beans are sacked or basketed and stored in the house.

[37] *Bixa orellana* was also widely cultivated for cosmetic purposes and wild plants were sought for the same use. For details *vide infra*, p. 118.

Vigna sinensis (*Cowpea*). Cowpeas are either planted to climb maize plants or trellises. Their treatment is the same as for beans. The cowpea is not widely cultivated among the Yukpa.

Cajanus cajan (*Pigeon pea*). The seeds are sown about 1.75 m. apart in April in soil which drains very rapidly. *Vípopa* and coarse *sásare* are especially favored for the plant. Although soil conditions indicate the gross conditions of the sowing place, sites near the house are most favored because of the plant's perennial nature. In the field it would quickly become lost in the rapidly growing fallow vegetation. The first harvest is made from eight to nine months after sowing, and although the plant continues to yield for three or four years, harvests decline sharply after the first.

Tephrosia singapou (*Barbasco*). Barbasco remains important as a fish suffocant. However, the plant is so widely encountered in accessible fallow areas that some consider the bother of cultivating it unmerited, especially now that the traditional fishing party has declined from a subsistence necessity to almost a ritual linking the tribe with the past. Changes in fishing technology are also playing a role; but suffocation remains the dominant, if diminished, technique of fishing. Barbasco continues to be planted in the shadier parts of gardens and fields. Seeds are sown at anytime during the year, and generally, after the initial sowing, the plants maintain themselves *in situ* by seed fall. Harvesting takes place year round, as materials are required.

EUPHORBIACEAE:

Ricinus communis (*Castor Beans*). The castor bean was once a vitally important crop of the Yukpa. Its seeds were gathered and used as an illuminant. The seeds were left on the plant until dry, and then gathered and stored inside the dwelling in a basket. When needed they were threaded on thin sticks of about one m. in length and used thus as torches. Recently, flashlights, candles, and electricity at the mission of El Tukuku have made its cultivation unnecessary, except among the less acculturated folk living in the most isolated communities. Some of the older Indians like to have castor beans growing near the house to provide shade, even though they no longer utilize the seeds.

Initially seeds were dibbled in new sites, always close to the house. Thereafter prolific propagation was by seed fall. In this state the plant reaches the size of a small tree, with a height of from five to seven meters.

Manihot dulcis (*Manioc*). Manioc is interplanted between the rows of corn, plantains, and bananas, and, as with all crops, is also grown in monocultural stands. It is spaced about 1.3 m. apart. Spacing is always very approximate and depends largely on the size and lie of unburned trunks, stumps, and rocks.

For planting, the soil is loosened to a depth of about 25 cm. with a *coa*. A small circular mound (*yayórwort*), about 7.5 cm. high and 15 cm. in diameter, is heaped up. From one to three cuttings are pushed at a slight angle into the mound and the loosened earth below it, covered over, and lightly tapped firm with the foot (pl. 12*a*). The cuttings may be either completely buried, or left partly exposed, depending on the whim of the planter. Care is taken to insure that they are not planted upside down.

Different varieties of manioc are grown together in the same field, but they are segregated in plots for the convenience of the harvester. Four different varieties with slightly different tastes are planted by the Maracá and Irapa (Irapa names in parentheses): *arášpoš* (*ķúwili*), a yellow stemmed variety, which yields well on poorer soils; *ķoķána* (*ķíhira*), distinguished by the whiteness of its stem, also yields very well on poorer, sandy soils; *mašríčaš* (*mašríčayi*), a much smaller plant, which is also characterized by having narrower leaves than the other varieties; and *šípa* (*šípa*), which is distinguished by having a very dark green stem and leaves.

Manioc interplanted in first year fields is not weeded until harvested. As each plant is harvested the surrounding area is thoroughly cleaned before the new cuttings are inserted. In second year fields it is, however, carefully weeded.

In harvesting, the area around the plant is first carefully cleared of weeds with a machete. The stems are then cut down to a height of about 15 cm. Standing with legs well spread, the woman grasps the stump and pulls firmly upwards to extract the tubers. Particularly compacted soil may first be carefully loosened with the machete. The earth in the hole is poked around casually with the machete, and mounded, and then cuttings from the previously removed stems are pushed back into the mound. Alternatively, the women may leave the 6 to 8 cuttings pushed into the hole for the man to plant properly a couple of days later. Manioc is planted only twice in the same hole.

SAPINDACEAE:

Cardiospermum halicacabum (*Heart Pea or Balloon Vine*). The cultivation of the heart pea has been discontinued except in the remotest communities, where it is sown both close to the house and, occasionally, in the fields. It is grown to provide thin vines for tying, and seeds for necklace making.

Seeds are sown throughout the year. Generally, after the initial sowing, the plants maintain themselves *in situ* by seed fall. Harvesting takes place as materials are required.

MALVACEAE:

Gossypium hirsutum (*Cotton*). Cotton is now a relict crop, and the occasional shrub is to be encountered only infrequently in older fields and close to house sites. Occasionally seeds are sown in April, but the majority of the plants are propagated by seed fall in fallow areas. The importance of cotton has been vastly reduced in the domestic economy as a result of the replacement of the traditional clothes by the dress of the *wátia* and bolts of commercial cloth which are made up into the tunic (*ķoštríča*) worn by men. Cotton is now used exclusively in arrow making, for which the amounts required are very small.

Hibiscus abelmoschus (*Hibiscus*). Hibiscus is sown close to the house, and occasionally in the fields. The fruit is used in ritual and as a medicine for curing snake bites. Except for the occasional ornamental plant in the gardens of the more acculturated Yukpa living in the mission at El Tukuku, the cultivation of hibiscus is continued only in the remotest communities.

Seeds are sown at any time during the year and generally, after the initial sowing, the plants maintain themselves by seed fall. Harvesting takes place throughout the year as materials are required.

BIXACEAE:

Bixa orellana (*Annatto*). Seeds of annatto are dibbled close to the house and the planting thereafter maintains itself by seed fall. Apparently it was never planted in field sites. The cultivation of this plant has been greatly reduced as the result of the decline in the frequency of celebrations necessitating the preparation of *sóña*. In times past the faces of all drinkers were painted with red dye obtained by mascerating the seeds of annatto in water. According to Clark (1938:1) red paint was also used to color the bones of exhumed bodies prior to the secondary burial. A minor present-day use of the annatto is for coloring cotton thread for decorative purposes.

CONVOLVULACEAE:

Ipomoea batatas (*Sweet Potato*). Sweet potato is usually planted in second year and third year fields, and close to the house site. Preferably sweet potatoes are planted on *sásare,* as the cutting requires a well-drained soil. More humid, clayey areas closer to stream courses are avoided as the crop is intolerant of waterlogging. Sweet potatoes are regarded by the Yukpa as a minor root crop, and planting therefore is undertaken after the more important crops have been taken care of. Replanting takes place throughout the year as tubers are harvested for immediate needs.

The plant is propagated from slips about 30 cm. in length, taken from the apical part of mature plants as they are harvested. The lower leaves of the cutting are removed and it is set at a slight angle into earth loosened with the *coa.* The cuttings are not consciously set out in rows, as sweet potatoes are generally used as a space filler. Rows are, however, occasionally made, in which case the cuttings are planted from 30 to 60 cm. apart. No further husbandry is undertaken with the exception of the general lackadaisical clearing of second and third year fields with a machete.

Harvesting takes place from July and August onwards as required for immediate usage. Storage is not practiced as the tubers deteriorate rapidly once removed from the soil.

SOLANACEAE:

Lycopersicon cerasiforme (*Tomato*). Tomatoes have been recently obtained from the *wátia,* and are only occasionally to be found. They are grown either close to the house or along trails in the fields, on any suitable vacant and accessible spot. Specific soil conditions are not a major requirement as tomatoes are very tolerant. They are regarded as a sweet fruit and are eaten only as snacks.

Propagation is generally by fruit fall from the parent plant, but occasionally seeds saved from the previous harvest are sown by broadcasting. Initial sowing takes place in April, and the fruit is ready for eating in August.

Capsicum frutescens (*Red Pepper*). Red peppers are planted near the house and also along trails in the fields. Any well-drained patch of soil is selected for sowing. Good drainage is important as the plants will not tolerate even short periods of waterlogging, which cause leaf fall. Seeds are broadcast only in new sites in March and April. The shrub is perennial and lives for two to three years and is therefore abandoned when the field is fallowed. Patches of peppers around the house site maintain themselves by fruit fall. Some of the fruits are harvested when ripe and are suspended from the house rafters to dry. Dry red peppers suspended from the rafters also serve to ward off malevolent spirits. Burning the fruit, however, is more effective. The plant has ritual uses only. Although it is never eaten, it is sometimes, but rarely, used to ease toothache.

Nicotiana tabacum (*Tobacco*). Every work group plants tobacco, for the Yukpa, from the youngest child to the oldest adult, are avid and incessant pipe smokers. It is sown in April after the start of the rainy season. Seeds are placed on the palm of the hand and, with a sharp puff of breath, are broadcast over the area to be sown. Level land is preferred, or if this is not conveniently available gentle slopes with plenty of large stones, behind which soil wash forms terracettes, are selected. Such sites are preferred because the light seeds are less easily washed away. In addition, the soil in which tobacco is sown should be moisture retentive, yet well drained, as the plant will not tolerate waterlogging. Preferred sowing sites are around the house, and along trails in the field. Such easily accessible sites are chosen because the care bestowed upon tobacco is unique among the crops of the Yukpa. Around each plant an area of several square meters is zealously weeded. At all times during the growing season of from four to five months, the plants are maintained scrupulously weed free.

Harvesting is done in August and September. The leaves are threaded on a twine made from *sápara* (*Agave viridis*) and suspended inside the house to dry. When perfectly dry the leaves are packed tightly together into long rolls, tied with any thin vine, and then stored by hanging them from the driest part of the house roof.

Datura candida (*Datura*). Every household has access to this hallucinogenic plant which is commonly encountered close to the dwellings. Among the Yukpa, *D. candida,* a small tree which reaches a height of from two to four m., is propagated exclusively by cuttings. The cuttings are planted concurrently with the field crops.

The Yukpa cultivate *Datura* solely for the psychotropic properties of the alkaloids present in the genus. The mind-affecting quality of this plant is put to shamanistic, specific, and general social uses. Unfortunately, I do not possess adequate details concerning its use in the shamanistic ritual.

Specifically, the use of *Datura* is thought to speed the growth of children. Flowers are picked by adults and filled with cold water, which mixes with the pollen. This concoction, which is said to be comparatively mild, is then drunk by the children who become quite intoxicated. Their drunken state may last for several days, during which time they often report frequent and unpleasant hallucinations.[38] Stronger potions are made from an infusion of the leaves of the plant. The infusion is then mixed with an enemy's *chicha* in order to render him unconscious. When in a coma the unfortunate is killed with arrows.

Infusions of *Datura* are used occasionally by the young adult males as a part of their general social life, a practice which occasionally results in riotous behavior and brawls. These infusions may be either the same as the mild

[38] For comparable reports from the Sibundoy region see Bristol (1969).

mixture taken by the children or the less frequently used and more powerful infusions made from the leaves of the plant. Under the influence of *Datura,* users claim to experience at first hallucinations of vividly beautiful colors and exquisite flowers and trees. Later in their intoxication they are terrorized by imaginary beasts of the forest. Apparently the hallucinations become increasingly frightful as the stupor progresses. In an effort to escape from these fictitious predatory animals many users shed their clothes and run naked through the woods, where they may become hopelessly lost until the drunkenness wears off.

RUBIACEAE:

Coffea arabica (*Coffee*). Production is most widespread among the Venezuelan Yukpa. Among the Maracá it has only recently been introduced to mission lands, where at present one Indian has one plant growing near his house. Two Indians are growing coffee on a field scale in the hot, low-lying Valle de los Motilones south of the mission at El Tukuku. One has chosen to interplant rows of bushes with plantains and bananas for shade, whereas the other uses the commoner method of retaining sufficient shade trees when clearing the forest to plant coffee. A number of Yukpa living in the middle and upper courses of the Tukuku and Negro rivers, plant coffee. It is commonest to clear a separate field for this crop, and interplanting is rare.

BIGNONIACEAE:

Crescentia cujete (*Calabash*). The calabash tree is cultivated by those groups of Yukpa which inhabit the *t. caliente.* The tree is generally planted close to the house. Although this small tree can be propagated from cuttings, the Yukpa use only seeds for this purpose. For uses *vide infra* in section dealing with *Laganaria siceraria.*

CUCURBITACEAE:

Cucurbita moschata (*Squash*). These are planted on lighter and well-drained soils, either in the fields or around the house sites in April. Spacing is irregular, but where plenty are grown it varies from 2.5 to 3.5 m. between rows of other plants. Mounds are not made and no particular attention is paid to husbandry, apart from the general clearing of the field. Harvesting of the ripe fruit takes place in August and September.

Lagenaria siceraria (*Gourds*). The naming of gourds among the Yukpa is at first sight somewhat confusing. They are designated by shape (and therefore by use) rather than for botanical identification. The term *"wíto,"* which refers without qualification to both *Crescentia cujete* and *Lagenaria siceraria,* is applied to near spherical gourds, the woody outer pericarp of which is used to manufacture bowls and servers. *"Pšášk̦"* is applied to more nearly bottle shaped gourds, which are of a variety of sizes and whose pericarp is used for transporting and storing water. The word *"sorúmspa"* is reserved for very large and thick-walled spherical gourds, which are used in the preparation of *sóña.*

Gourds are sown both in the fields and around house sites in March and April. They are ready for harvesting in August and September, although some are harvested when small as early as June and July. The time of harvesting depends on the sizes of utensils required. Sizes of gourds are also, among the more sophisticated tribesmen, determined by seed selection. Sowing is haphazard and gourds are used to fill in space rather than for interplanting, *sensu stricto*.

Two alternative procedures are followed in harvesting: According to the first, gourds may be harvested when dead ripe. They are left to dry out thoroughly on the vine, then the top is cut off and the flesh and seeds emptied out. According to the second procedure, the gourds may be picked when still green, and the flesh and seeds then routed out with a pointed stick. The woody outer pericarp is left in the sun to dry out for several days. This is generally not the procedure followed for *pšášķ*. One can readily appreciate that this would be more difficult with narrow and often curved neck shapes.

ARBORICULTURE:

Tree crops, especially citrus, and coffee, which have been obtained from the *wátia,* are now of increasing importance and are slowly spreading to the more isolated communities. Apart from a few cultivators who are now beginning to grow coffee on a field scale, the newly introduced tree crops generally occur as isolated specimens around the house site. Occasionally a single tree may be encountered in a field site, but this is a rare occurrence.

A range of fruit trees has been adopted from the *wátia,* especially by the Yukpa of Venezuela. Fruit remain of little importance, however, in the diet of the Indians. Cultivated trees are mostly grouped around the house site, where they are relished for the shade which they provide and where newly fallen fruit may be conveniently gathered for use. The mango (*Mangifera indica*) is particularly favored as a shade tree. Only occasionally are fruit trees planted in the field sites as nobody ever bothers to look in fallow vegetation when harvesting fruits.

The commonest of the adopted trees are Rutaceae (*Citrus aurantfolia,* lime; *C. paradisi,* grapefruit; and *C. sinensis,* sweet orange), followed by the mango (*Mangifera indica*), papaya (*Carica papaya*) and the sapodilla (*Achras zapota*). Occasionally guanabana or soursop (*Annona muricata*), breadfruit (*Artocarpus altilis*), genip (*Melicocca bijuga*), guava (*Psidium guajava*) and the mammey apple (*Mammea americana*) are to be found. One cultivator recently planted a coconut tree (*Cocos nucifera*) after he was delighted by the taste of the fruit during a visit to Machiques, and where he purchased a young plant.

The inventory of crops grown by the Yukpa is changing rapidly. New species, new varieties, and plants hitherto unknown are now being planted experimentally, and on an ever larger scale. Along with these new plants have come changes in the structure and the technology of the cultivation system. Some innovations are of course rejected, but those which can be readily integrated into existing patterns and which can be seen to be beneficial to Yukpa life are eagerly accepted. Changes are not always positive, and concomitant with the acceptance of the new comes an abandonment of the old. As old techniques fell into disuse some plants were no longer cultivated, but the staple core of Yukpa crops is unlikely to change in the immediate future. Around that core is emerging a complex pattern of acceptance of new crops, and the abandonment of other fundamentally less important crops.

PHASE FIVE: FALLOWING

The Necessity of Fallowing

The final phase of the Yukpa system of shifting cultivation is fallowing. Fields are cycled out of cultivation after one, two, and three or more years; and, either the natural vegetation associations are permitted to return, or, infrequently, they are planted with pasture grasses.

The Yukpa recognized that the fallowing is necessary for the following reasons: first, to replenish the store of soil nutrients which are gradually removed from the soil when forest is cleared and the land cropped; second, to reconstitute desirable soil structure; and third, to eliminate weeds. During the cultivation cycle, and particularly during second year and later usage, the increment of weeds, especially stoloniferous grasses, reaches such proportions that yields of annual crops are drastically reduced. Under semi-perennials, such as yautia, however, the problem of weed infestation is significantly reduced. The Yukpa are not able to eliminate weeds, particularly grasses, without an excessive input of labor. The system of shifting cultivation from the viewpoint of the input-output ratio, is more efficiently operated by clearing new fields, and permitting forest growth to recover the old field. Under the tall fallow vegetation, the noxious grasses are effectively shaded out.

A fourth use of fallowing recognized by the Yukpa is that it helps eliminate crop diseases and animal pests. In a cultivated field, a few plant species are concentrated in large numbers within a small area, and concomitant with this is the concentration of crop pests and diseases, and hungry animals which reduce crop yields (Bowden, 1956; and Cutting, et al., 1957, cited by Nye and Greenland, 1965:76; Crist, 1969:256). Fallowing permits the reclamation of land by a greater diversity of plant species, thus eliminating the concentration of pests. Shifting cultivators in the area have an advantage over the agriculturalists of the Maracaibo Lowlands and the César Valley, in that they preserve the crop pests' natural enemies closer to their pre-agricultural level (Labrador, 1967:61).

Factors Influencing the Duration of the Fallow

Two universal factors influence the duration of the fallow among shifting cultivators. The principal factor is the pressure of population on the economically usable land, and, associated with this, the distance of land suitable for cultivation from the pivotal settlements. The second factor is the time necessary to regenerate vegetation associations which the cultivators consider suitable for reclearance. The length of fallow can therefore vary widely depending on the type of soil, and the degree of impoverishment of the soil and vegetation by cultivation and other autosubsistence activities.

Population Pressure. Within the Yukpa territory as a whole there are no acute problems of population pressure deriving either from internal population growth or from the encroachment of *wátia* colonization. The broad picture is, however, complicated by a number of local details.

The division of the tribe into a number of traditionally warring subtribes, who now live in a somewhat uneasy peace with each other, virtually precludes the free movement of members of one subtribe into the territory of another for the purpose of land settlement. Thus the space available for cultivation is limited to subtribal territory only. This is complicated by the fact that a man living with one subtribe, because of the matrilocal residence rules, may have inherited land from his father in the territory of another subtribe. Subtribal territorial boundaries may therefore be crossed at the individual level owing to kinship and inheritance rules. Because of territoriality the possibility of some of the subtribes' experiencing population pressure could be acute were it not for the fact that the population density of all of the subtribes is very low.

The impact of *wátia* colonization is not felt equally on opposite sides of the international border. In the past the Yukpa were progressively pushed back toward the Sierra de Perijá in advance of the encroaching frontier of Venezuelan *haciendas.* The establishment of an Indian reserve, however, reduced this threat to the Yukpa livelihood to an insignificant level. Alienation of Yukpa lands still takes place within the reserve, especially in the Negro Valley, by *wátia* operating small *conucos,* but happily the number is very small.

On the Colombian side of the frontier, the situation is completely different. There is no official Indian reserve, and the Yuko subtribes have been reduced to a very small area, which is quite inadequate for their needs. The narrow valley in which the Maracá live, for example, is completely surrounded by the *fincas* of Colombian *campesinos.* Occupation is *de facto,* and the *wátia* can take the land at will.

In this region population pressure is a locally important factor which in certain communities forces a drastic reduction of the length of the fallow, with all of its disastrous ecological implications. It must be added, however, that if the Yukpa themselves caused less destruction by their extremely careless practices, the incipient population pressure upon the subtribes in Colombia would constitute considerably less of a problem.

Environmental Factors, Cultural Concepts, and the Duration of the Fallow. Environmental factors, including soil type and degree of vegetational impoverishment, together with the cultural concepts of type of vegetation required before a site can be cleared again play an important role in determining the length of the fallow. For the Yukpa the ideal type of vegetation for field sites, regardless of soil type, is high secondary forest fallow, 30 to 35 years of age. However, confining clearing to such vegetation would necessitate frequent house site shifting. Hence, field clearance most commonly takes place on sites with low forest fallow vegetation, 12 to 15 years of age, which are more conveniently located with reference to the settlements.

Such statements are only generalizations, because the soil type variable has to be considered. For fields cleared on *vípopa* soil, there is no fixed length of time for the fallow. Ideally they are fallowed for "as long as possible" between periods of cultivation, the actual time period depending on population pressure, the food needs of the work group, and the physical characteristics of the remainder of the individual's land. Fields cleared on *vípopa* are generally cultivated for one year only and then fallowed. Because of the repeated uncontrolled burning in the area, deflected successions, dominated by highly flammable ferns, are perpetuated. Thus the impoverishment cycle on *vípopa* is kept in motion. On areas of *sásare* soils the fallow period approaches ten to 12 years, whereas on black earth the length of fallow varies from two to 12 years, depending on the microecology of the site and the needs of the individual work group.

The length of the fallow in the Yukpa system is quite variable. Short and long fallow systems operate in close proximity to one another. One work group may operate both long and short fallows on black earth, long and short on *sásare* and long on *vípopa,* whereas the neighboring group may operate long on all soils, or short on all soils. There is little variation within long systems, where three or four years of cultivation is followed by from ten to fifteen years or more of fallow. Short systems exhibit greater variability: one or two years of cultivation and one or two years of fallow; one or two years of cultivation and five years of fallow are those most commonly used.

The question is complicated. The determinant variables are analyzed within the restraints of the individual work group's needs, capital, and land, both owned and borrowed.

Factors Affecting the Quality of the Fallow

The ideal of the Yukpa is to promote a plant sucesssion on evacuated field sites leading to the progessive re-establishment of a high forest fallow. In some localities this is accomplished, whereas in others it is precluded by a disastrous pauperization of the soil and vegetation.

The main factors affecting the quality of the fallow are: the type of vegetation at the site when cleared; the cultivation-fallow ratio; the surrounding secondary-growth vegetation; uncontrolled burning; and the other types of land use in the vicinity.

Types of Vegetation at Site When Cleared. The type of vegetation originally cleared plays a role in determining the type of succession which follows after the field is evacuated. If the field was originally cleared in high forest fallow the same associations tend to return rapidly with the end of cultivation because of the sprouting of stumps and the germination of seeds. The earlier stages of transient annuals and grasses and light-loving trees are shortened. Conversely, when a field has been cleared on *vípopa* in associations dominated by ferns, ferns and Gramineae tend to dominate the earlier stages of the succession, which, together with the impoverishment of the soil that permitted the initial dominance of fern, slows up the ingress of the later stages.

Cultivation-Fallow Ratio. If the period of cultivation is not so long that extensive pauperization of the soil takes place, and if the time permitted for fallowing is sufficient to allow an uninterrupted succession to high forest, the quality of the fallow, judged by all commonly accepted criteria, is high. If, however, the period of the fallow is greatly reduced before recultivation, poor quality secondary vegetation, dominated by Gramineae, may result in "deflected" successions during ensuing fallow periods.

Above all it must be remembered that the concept of "quality" is entirely relative and is culturally determined. Good quality fallow vegetation to the Yukpa is that which is sufficient to permit the return of the ecological conditions required for the production of a good harvest. Yet, traditionally, the long-term ecological implications of their subsistence activities seem to have been little considered. Thus they justify the foreshortening of the cultivation-fallow ratio in terms of immediately succeeding crops yields. It should therefore be noted that the Yukpa concepts of "good quality fallow" and "ideal fallow" are not necessarily coincident. This dichotomy has to be seen in the light of prevailing immediate conditions.

Surrounding Second-Growth Vegetation. The quality of fallow vegetation is in large part controlled by the nature of the nearby vegetation associations. Wind-blown and faunally carried seeds, and grass root systems spread into the evacuated field from surrounding areas. Thus, the light, readily distributed seeds of grasses and ferns, together with light-loving secondary forest associations, rapidly colonize field plots given over to fallow.

Only a minority of the Maracás' fields are completely surrounded by high secondary forest associations. The majority have at least one boundary with savanna and fern dominated areas. Thus, the normal percentage of grasses and annuals in the early stages of secondary succession is reinforced, particularly if the previous fallow periods were greatly reduced.

In other localities, especially in the lower Tukuku and Negro valleys, where population pressure is not so great, and the longer period of contact with the *wátia* has resulted in the adoption of better husbandry techniques, areas of "deflected" succession are much smaller. It is rare to find plots bounded by grassland, and most plots are completely surrounded by a mosaic of high forest fallow and *rastrojo.* Thus, the ingress of noxious grasses in evacuated fields is greatly reduced, and conversely the fallow vegetation is quickly dominated by ligneous growth.

Uncontrolled Burning. By far the most important single factor affecting the quality of vegetation in fallow areas is the careless use of fire. Areas in fallow are not protected, and repeated burning of timber takes place within them, especially those located on the steep slopes (pl. 13) as a result of both field clearance activities and the noncultivation uses of fire.

Other Types of Land Use in the Vicinity. Localized areas of savanna influence the fallow areas not only by supplying seed and rhizomes, but also by fueling escaped burns which spread into fallow areas. With the *wátia,* however, came livestock, pastures and permanent field crops, such as coffee. In some instances several years of cultivation ended with the planting of pasture grasses, maintained with fire. Planted pastures are mostly limited to small areas around the missions at Socorpa and El Tukuku, and the lower Negro Valley. Indians living in the remoter areas generally provide poor quality forage for their livestock by maintaining the savannas, with the help of annual burning.

Treatment of the Fallow Areas

After a period of time which ranges from one to four years, providing pasture grasses are not sown, fields are evacuated for fallowing, the final stage of the cultivation cycle. Treatment of the *rastrojo* areas may be characterized most effectively in a negative manner; for, apart from use as relatively rich hunting preserves, and as a minor source of wild plants (which may be gathered more easily elsewhere [Ruddle, 1970*a*:264–269]), the fields in fallow are neither treated to favor specific plant successions, nor are they protected from escaped burns. Although wild cane is planted on the humid stream course margins of fallow areas nothing is done to control or to enrich the regeneration.

Perennial and semi-perennial crops remaining in the regenerating fields, together with edible wild plants which form an integral component of the regressing natural vegetation, may be collected until such time as access becomes too difficult. This is not usually done, however, except in cases where no other food is available, because the remaining crops are generally damaged by faunal activities. In any case, by this time other fields which are more convenient of access and yielding better quality crops are in production. It is just not worth the effort to search for crops in the *rastrojo.*

During the period when field products are scarce, however, wild roots, tubers and shoots become correspondingly more important in the daily diet, but only in the most severe cases of food shortage do they assume a dominant position. It should be remembered that families with food are obliged to feed those without. So, unless there is a general food shortage, the hapless prefer to ask for food rather than search for the less relished wild products. In periods of food shortage, however, wild roots and tubers are sought in the cultivated fields, in regressing vegetation and in the *rastrojo.*

The collection of roots and tubers, which was once a vitally important Yukpa subsistence activity, is now virtually extinct. In times past, according to both oral history and folk literature,[39] a wide range of what are now regarded as wild plants were cultivated in the same manner as the contemporary crops are handled. According to legend, however, these old crops were rejected when the modern crops were obtained. Unfortunately, these ancient cultivates, which my informants say were "grown in rows like *míšk* and *póyši,*" are now extremely difficult to locate in their feral state. After three field seasons I have still been unable to procure specimens of four of the plants for identification. Former cultivates of this tribe include: *čuña* (*Dioscorea placidae*), *čipéntre* (*Dioscorea* sp.), *kamása* (*Carica microcarpa?*), *komitrúšk* (*Physalis peruviana*), *nopášwots* (*Bomarea granatensis*), *oyášk* (*Solanum nigrum*), *pásia,*[40] *pirínay,*[40] *sakóra,*[40] *sáata,*[40] *šiapa* (*Gustavia angusta*) and *ɨdkɨšk* (*Chamaedorea* sp.). Wilbert,[41] when working among the Parirí group in 1959, learned of an edible root known as *makáhka,* which may also have been cultivated in former times. Regrettably, no specimens of this plant, either, could be found for herbarium identification.

The cultivated field and the *rastrojo* also furnish a wide range of edible wild plant products (table 4). Forest products are, of course, only available at certain seasons. Fortunately, the slack period of the cultivation cycle, when gathering assumes greater importance, coincides with the major harvest season of the forest fruits.

It is extremely difficult to estimate that percentage of the diet comprised of wild plant products. Wild plant leaves are generally consumed as green vegetables two or three times per week throughout the year. They comprise about 60 percent of the greens consumed. The other 40 percent of green vegetables consists of leaves of taro, yautia and squash.[42]

The most important alimentary role of wild vegetation is the production in the high forest fallow of fruits which are eaten raw as snacks. Among

[39] Wilbert, 1974.

[40] No specimen procured for identification.

[41] Personal communication.

[42] Wilbert (1960:109) noted that among the Parirí leaves of cultivated plants are consumed mainly by women and rarely by men.

TABLE 4
WILD FOOD PLANTS GATHERED FROM CULTIVATED FIELD, RASTROJO, AND FOREST SITES

Yukpa	Botanical identification	Part used[1]	Preparation
Anápa	*Ficus* sp.	Se.	Eaten raw
Kátša	*Inga* sp.	Se.	Eaten raw
Kɨḵia	*Lantana camara*	Fr.	Eaten raw
Kɨnɨri	*Solanum hirtum*	Fr.	Eaten raw
Konónopira	*Psidium guineense*	Fr.	Eaten raw
Kuwiwiḵan	*Inga spuria*	P.	Eaten raw
Mapɨtḵa	*Inga* sp.	R.	Eaten raw
Mátlia	*Clavija longifolia*	Fr.	Eaten raw
Oónikara	*Cissus rhombifolia*	Se.	Eaten raw
Šámpaša	*Piper pavoni*	Fr.	Eaten raw
Šára	*Poulsenia armata*	Fr.	Eaten raw and as a beverage
Tamáḵrašḵ	*Acacia* sp.	L.	Boiled and eaten raw
Tohíḵu	*Piper taboganum*	Se.	Eaten raw
Tútra	*Rollinia edulis*	Fr.	Eaten raw
Vuháḵa	*Maranta arundinacea*	R.	Boiled/baked

[1] Symbols for "Part Used": Fr = *F*ruit, L = *L*eaf, P = *P*ulp (the white pulp around the seeds in the pod), Se = *S*eed.

the majority of the Indians approximately 90 percent of the fruit snacks consumed are obtained from the forest. In communities close to *wátia* settlements forest fruits are being increasingly displaced by cultivated products. Even in such acculturated communities, however, cultivated fruits have not yet attained a level of production sufficient to relegate wild fruits to memory only. Indeed, it is highly unlikely that this will occur within the foreseeable future, for the wild products provide a range of tastes and a variety unequaled by the cultivated fruits.

The most important of the fruits gathered by the Yukpa is the *šára* (*Poulsenia armata*), the harvest of which constitutes a major event in the tribal calendar. No other single forest fruit merits the Yukpa's attention to such a degree. *Šára* is collected in May and June, at which time all of the subtribal members, with the exception of the oldest and the infirm, descend to those places in the lowlands where the fruit grows.

Although the actual labor of harvesting is done by the members of the individual nuclear families, the whole community goes together to the harvesting sites, to ensure fair sharing (the trees are communal property), and, more important, especially in the recent past, for protection against the *wátia*. The difficult and dangerous work of fruit picking is done by the men, who construct a ladder (*šára tápoymotos*) of saplings (*yápoymor*) and lianes (*mimíšk̜*) enabling them to climb to the highest branches of the trees. The men toss the fruits down to the waiting women and children. This is the normal division of labor for harvesting the forest tree crops for which the general collection procedures are similar. The bulk of the harvest is eaten at the temporary camp close to the harvest site. *Šára* is consumed in three ways. Frequently it is eaten raw, but more often the fruits are boiled in water and eaten as a hot dish. The fruit is also ground up to make a cold, refreshing early morning drink (*šára tɨrk̜ás*).

Šára is a delectable fruit. Its harvest is a pleasant interlude in the yearly round, when the Yukpa community can relax after the arduous task of preparing and planting the new season's cultivated field.

The *rastrojo* and cultivated fields are also important as sources of medicinal, magico-religious, ornamental, technological, and poisoning plants. The importance of the tall forest and savanna areas, however, far exceeds that of the *rastrojo* and cultivated field as a source of plants for those purposes (Ruddle, 1970*a*:335–361).

The tall forest fallow is, quantitatively, the most important source of medicinal plants. From this biome 94 of the useful plants (51 percent of the total named by informants) were derived, compared with 48 (26 percent) from the savanna, 31 (17 percent) from the *rastrojo* and only 11 (6 percent) from the cultivated field. From the cultivated field and *rastrojo,* the medicinal plants listed in table 5 are obtained.

The cultivated field and areas of *rastrojo* are also a minor source of plants used for magico-religious purposes. Only three plants, *hápa* and *potídčo* (*Rynchospora* sp. and *Bidens pilosa,* respectively), the leaves of which, when used to prepare an infusion for bathing in, are thought to give good aim to hunters, and *k̜aráša* (*Renealmia occidentalis*), whose seeds are used to make necklaces which protect against evil spirits, are gathered in the cultivated field. Areas of *rastrojo* provide only two plants with magico-

TABLE 5

MEDICINAL PLANTS OBTAINED FROM FIELDS AND RASTROJO[1]

Yukpa	Botanical identification	Part[2] used	Preparation	Purpose
		CULTIVATED FIELDS		
Gotúmo	*Phytolacca octandra*	Fr.	Eaten raw	Combats *"Carate."*[3]
Hápa	*Rynchospora* sp.	Fr.	Squeezed in cold water	Washing cures boils.
Huántu	*Setaria paniculifera*	L.	Infusion:[4] drunk	Alleviates fatigue.
Kɨḵia	*Lantana lilaeina*	Fr.	Eaten raw	Cures general "queasiness."
Kómar ḵátača	*Justicia secunda*	Fl.	None	Rubbed over body to cure sunstroke.
Kunpánḵa	*Cordia ferruginea*	L.	Infusion: bathed in and drunk	Cures sunstroke and headache.
Kurása	*Renealmia occidentalis*	L.	Infusion: bathed in	General panacea, especially for rheumatic pains.
Matánon biyáiya	*Dichronema ciliata*	L.	Infusion: drunk	Cures hangover and the "dry heaves."
Pátubia	*Leptochloa virgata*	L.	Infusion: drunk	Cures stomach-ache.
Taḵíišu	*Melochia globifera*	L.	(a) Infusion: bathed in (b) Poultice of leaves placed over child's anus.	Alleviates sunstroke and reduces fever. Laxative for children.

TABLE 5 cont.

MEDICINAL PLANTS OBTAINED FROM FIELDS AND RASTROJO (cont.)

Yukpa	Botanical identification	Part[a] used	Preparation	Purpose
		CULTIVATED FIELDS		
Wašówrača	*Erechtites hieraceifolia*	L.	Infusion: bathed in	Reduces fever.
		RASTROJO		
Hóya	*Solanum* sp.	L.	Infusion: drunk	Cures stomach-ache.
Kɨkia	(see above, in "Cultivated Fields")			
Kiníyapa	*Valeriana pavonii*	L.	(a) Infusion: bathed in (b) Ashes mixed with water and drunk	Reduces fever.
Mápola	*Lycoseris oblongifolia*	L.	Infusion: drunk	Stops women from hemorrhaging.
Mátuba	*Solanum* sp.	Fr.	Applied directly	Cures *"Curate."*[3]
Mi biyáiya	{ *Echinochloa crus-pavonis* *Hackelochloa granularis* *Heteropogon contortus*	{ L. L. L.	{ Infusion: drunk or bathed in	Cures hangover after *míšk* and *sóña* orgies.
Ñame biyáiya	*Dioscorea* sp.	L.	Infusion: drunk	Cures illness caused by eating bad yams.
Oóhwia biáiya	*Catasetum* sp.	L.	Infusion: bathed in	Cures indigestion.
Otéiya	*Commelina* sp.	L	Infusion: bathed in	Reduces fever.
Ówa biáiya	*Lindsaya guianensis* (?)	Sa.	Drunk	Contraceptive for women.

TABLE 5 cont.

MEDICINAL PLANTS OBTAINED FROM FIELDS AND RASTROJO (cont.)

Yukpa	Botanical identification	Part[2] used	Preparation	Purpose
Sarumba	*Eleocharis fistulosa*	St.	Cut to size	Splints.
Séntre	*Relbunium hypocarpium*	L.	Infusion: bathed in	Alleviates fatigue.
Tiími kɨríka	*Iresine acicularis*	L.	Infusion: drunk	Laxative.
Uunéi kára	*Cissus thombifolia*	L.	Infusion: bathed in	Cures pain in joints.
Wápia	*Dryopteris patens*	L.	Infusion: bathed in	Panacea.

[1] Only those plants for which specimens were procured for herbarium identification are listed.
[2] Symbols for "Part Used"; Fl = *Fl*ower, Fr = *Fr*uit, L = *L*eaf, Sa = *Sa*p, St = *St*em.
[3] *Treponema carateum.*
[4] Infusions are always prepared with cold water.

religious uses: *ersók̠ano,*[43] the leaves of which, when bathed in by a woman, are considered to change a hunter's bad luck; and *watɨ́k̠wan* (*Wedelia fruticosa*), the leaves of which, when rubbed on a hunter's face, are believed to improve his aim.

Some plants used for personal adornment are also obtained from the cultivated field and *rastrojo.* Of these the most commonly collected are *k̠imímbri* (*Desmodium sericophyllum*), *k̠ótomo* (*Phytolacca icosandra* and *P. rivinoides*) the fruit of which yields a facial paint for ceremonial purposes and dye for textiles and basketry. From the fallen logs and stumps of the cultivated field, *rastrojo* and the high forest fallow a minor yet nevertheless important commodity is collected. This is the group of fungi used by the Yukpa for poisoning their enemies. Of these the most important is the ubiquitous *owák̠yuk̠a* (*Polyporus* sp.) which give rapid results.

The fallow is used as a source of firewood and construction timber. Standing trees killed by escaped burns, and felled trunks not completely consumed by burning during the cultivation cycle, are utilized for firewood. The denser wood of slow-growing, later stage secondary growth vegetation is favored for firewood. Fast-growing, light-loving trees such as *širíyi* (*Cecropia* sp.) are never used for this purpose, but instead they are cut for construction and fencing timber.

Apart from the *Cecropia* sp., the cultivated field and areas of *rastrojo* are of only minor importance for the collection of plants used in Yukpa technology. In field clearance, technologically important plants are not purposely destroyed, but neither are they protected during firing, nor are they paid any special attention during the cultivation cycle.

Occasionally encountered in these sites are the *nɨ́rpɨtɨ* (*Furcraea gigantea*) the scape of which is used for making quivers (Ruddle, 1970*b*:24–25); *puuk̠* (*Stromanthe lutea*) used for the weaving of baskets, fans, hats, mats, and quivers among other things; *téva* (*Guilielma* sp.) which provides wood for bows; *sapára* (*Agave viridis*) used for rope and string manufacture; *wóši* (generic name for grasses) and *toróma* (*Heliconia bihai*) used for thatching.

Plant Succession

On fallowed land in this region two types of successions are to be found: the desired, "progressive" succession leading eventually, if uninterrupted, to the re-establishment of a forest climax; and the undesirable, "regressive"

[43] No specimen procured for identification.

or deflective succession dominated by grasses or ferns. An analysis of the regressive successions has been presented in Chapter 2. Only the successional stages leading to the desirable high forest secondary climax are outlined here.

Succession to Secondary Forest. On evacuated fields which escape accidental burns, three stages leading to the re-establishment of climax forest may be noted. The beginning of the fallow phase is not normally marked by even a fleeting period when the evacuated field is relatively bare. Occasional exceptions to this are cases where a first year field, cut in high forest fallow, which generally results in a fairly weed-free plot, is evacuated. Not only are a few perennial crops such as plantains, bananas and sub-perennial root crops remaining in the field, but what is more, the field is also fairly densely covered with occasional surviving palms and regressing natural vegetation.

As a result, there is no really clear-cut demarcation between the cropping and fallow passes of the cultivation cycle. If a field is evacuated after only a year of *míšk̨* cultivation then it may be said, from the viewpoint of the plant associations regressing, that the commencement of cropping and fallowing is coincident, because the field was never weeded. Second and later year fields dominated by root crops are only haphazardly cleaned once or twice during the growing season. Thus the herbaceous weeds, which mark the first stage of the succession, are well established at the end of the final cultivation cycle. Because of the great variability of the Yukpa cropping to fallow ratio, the character of the first stage of the "progressive succession" may vary from place to place in terms of its components. On fields which have been cultivated for one year only, the percentage of herbaceous weeds, especially Gramineae, is low, whereas that of resprouting tree stumps (*pánatarik̨*), and the light-loving trees (especially *Cecropia* sp.) is high. In fields which have been cleaned and burned during succesive years of cultivation the remnantal ligneous material (especially stumps) may have been almost completely killed off, and the percentage of lower herbaceous growth is greater.

Regardless of its duration, the initial successional stage is dominated by low herbaceous growth, mainly comprising annual dicotyledons and Gramineae. This first stage of the succession is dominated by: Euphorbiaceae, (particularly *Euphorbia elliptica, geniculata, heterophylla,* and *lasiocarpa*), Solanaceae (especially *Solanum nigrum, hirtum* and *scabridum* and *Physalis peruviana*), Verbenaceae (mainly *Lantana camara* and *lilacina, Coccocypselum lanceolatum, Manettia calycosa* and *Relbunium hypocarpium*), and Compositae (particularly *Ageratum conyzoides, Bidens pilosa, Erechtites hieraceifolia, Eupatorium microstemon, Wedelia fruticosa* and *Galinsoga* sp.). A smaller percentage of the first stage herbaceous growth is composed of Malvaceae (especially *Sida acuta* and *glomerata*).

The Gramineae present in the first stages of succession comprise both annuals and perennials, although the latter form the majority. The species of Gramineae most commonly encountered include: *Andropogon* (*bicornis**, *leucostachys**, and *saccharoides**), *Aristida recurvata**, *Echinochloa crus-pavonis, Eleusine indica, Eragrostis compacta, Eriochloa distachya**, *Gymnopogon spicatus, Hackelochloa granularis, Heteropogon contortus**, *Hyparrhenia bracteata**, and *rufa**, *Lasiacis procerrima**, *sloanei**, *sorghoidea**, *Leptochloa domingensis**, *Melinis minutiflora**, *Panicum glutinosum**, *olyroides**, *rudgei**, *sellowii**, *Paspalum canescens**, *heterotrichum**, *plicatulum**, *Pennisetum bambusiforme**, *setosum**, *Setaria paniculifera**, *poiretiana**, *vuipiseta**, and *Trachypogon plumosus**.

During the second year of the succession, which is now about 2 m. high and of increasing impenetrability because of vines and climbers, the annuals die off, leaving the coarser perennial grasses (denoted by asterisks in the above list) to assume greater importance. At this point the larger-leafed monocotyledons, especially *Heliconia bihai, Costus spiralis, Maranta arundinacea,* and *Stromanthe lutea* become more noticeable. These gradually shade out the smaller annuals, and coexist with the more persistent perennial grasses.

In this stage, too, a number of coarse composites enter the succession, particularly, *Verbesina columbiana, Clibadium surinamense, Baccharis trinervis,* and *Vernonia brasiliana,* together with the light-loving, very rapidly growing trees and bushes such as *Cercropia* sp., *Cordia alliodora, curassavica, ferruginea,* and *polycephala,* and *Heliocarpus americanus* and *popayanensis,* which at this time begin to form the upper level of the succession. Gradually these light-loving trees, together with the broad-leafed herbs begin to dominate the early years of the succession. If the area is unprotected from fire, and the plant succession continues uninterrupted, the bulk of the more persistent perennial grasses is shaded out.

Gradually, as the lower growth is shaded out, slower-growing trees and bushes which characterize the later stage of the succession become more widespread. Particularly common are *Miconia* spp. (notably *albicans, serrulata, theaezans*) together with plants of *Ficus* and *Inga* genera. Eventually, in about 15 to 18 years, if the succession is not interrupted the species characteristic of the early stage of the succession are replaced by a host of forest species.

The secondary forest at about 20 years of age is a complex association of trees, bushes, grasses, lianes, and herbs of various sizes. The ground is covered by a thick, almost impenetrable underbrush. Eventually, given continued protection of the area, the vegetation begins to assume the characteristics of the primary forest.

Chapter 5

LAND USE, LABOR, AND PRODUCTIVITY

The annual cycle of cultivation has been analyzed. In this chapter the spatial and temporal aspects of the cultural landscape created by the functioning of the autosubsistence system, the labor input required to operate it, and its productivity will be examined.

Techniques in Continued Cultivation

In addition to clearing a new field each year, the Yukpa cultivator has to prepare the previous year's fields for further use. In some first and second year fields, *rastrojo* is permitted to return; but in others, which are to be replanted, some form of preparation is undertaken at the beginning of the second and third year of cultivation.

New fields which were cut in high forest fallow are not generally weeded during the first year of cultivation. First year fields cut in *rastrojo,* however, become heavily weed infested during the cropping period so that occasional weed cutting is necessary.

After the first year crop has been harvested the field is left completely exposed to the elements and to weed infestation. If the weed increment is slight, and if the soils are friable, substantial soil erosion takes place during the last few months of the rainy season.

During February of the second year the weed increment and resprouted stumps are slashed and the material permitted to dry out. If the weed growth during the time elapsed since harvesting was slight, and if the heavy ligneous growth originally growing on the site was almost entirely consumed by the first year's burning, the field is not burned before being reused. Instead the leafy matter is left to dry and mulch, because it is realized that the benefits of burning a small amount of slash would be outweighed by the increase in soil wash potential resulting from the greater friability of the soil induced by a second burn. Further, sections of the first year fields which were planted in perennial and semi-perennial crops such as plantains and bananas, sugar cane, and manioc are cleaned and not burned. If, however, the weed increment was substantial during the first year, and especially if the first year burn was poor, fields are burned before being replanted.

Using the sowing and planting methods already described, the second year field is planted. The main part of the field is progressively planted in manioc over a period of several months depending upon the availability of suitable cuttings, and the desire to stagger the harvest. Widely spaced rows of plantains and bananas are then planted in the most accessible part of the manioc field. Occasionally maize may be sown, but this is only rarely done as the yields are reduced by about 50 percent in the second year of use.

The increment of weeds in second year fields is rapid and substantial, especially in fields originally cut in *rastrojo*. Accordingly, the manioc field is cleaned by cutting the weeds with a machete about once every two months when the plants are young. The dense leaf canopy of older, closely planted manioc effectively slows down the increment of weeds, such that the interval between cleanings is lengthened.

Plantains and bananas planted in second year fields are weeded only once a year, and the overgrown and weed infested appearance of young plantations suggests return to fallow. In reality this is not so, as the older plants are gradually able to shade out most of the weeds. Weeds cut in the manioc and plantains and bananas are either left to mulch where they fall, or, more rarely, are piled in heaps to one side of the field where they may be burned without damaging the standing crops.

At the end of the second year of use, as the manioc is gradually harvested, fallow vegetation is permitted to return to all but the continuing plantain and banana units. The plantains and bananas are given once yearly weeding so long as they continue to yield sufficiently well. Some fields appear to contain plant clusters of about ten years of age. As will be illustrated below in the case studies, where second year fields are located adjacent to the house, sections are often diversified by planting perennial crops at the beginning of the third year. In this manner parts of some old fields are converted into dooryard gardens.

The Sequence of Field Use

Yukpa land use presents, at first sight, a bewildering variety of crop combinations, field use sequences, length of field use, and dispersal of field types. The combinations of crops planted in a particular field vary according to the whim and taste of the cultivator and his family, the availability of seed and plant parts for planting, and the size and ecological characteristics of the field. Field use sequences vary, depending largely on the changing physical characteristics of the field and on a large number of disruptive sociocultural factors, such as illness, death, strife, loan of land, and cooperative cultivation. Despite a wide range of variations and deviations from the system, the underlying, traditional or ideal land use sequence and associated field-use types may be discerned.

TABLE 6

YUKPA CROP SEQUENCES: TRADITIONAL (IDEAL) SEQUENCE

Season 1	Season 2	Season 3	Season 4	Season 5
MAIZE[1]	MANIOC	PLANTAINS & BANANAS	PLANTAINS & BANANAS	PLANTAINS & BANANAS
Beans	Plantains & bananas			
Squash				
Gourds				
Pigeon peas →	Pigeon peas			
Other annuals	Squash			
	Gourds			
Manioc →	Manioc			
Yautia →	Yautia			
Sugar cane →	Sugar cane →	Sugar cane		
Plantains & bananas →	Plantains & bananas →	Plantains & bananas →	Plantains & bananas →	Plantains & bananas
Millet →	Millet			
		Fallow →	Fallow →	Fallow

[1] The dominant crop, which defines the field, is in capital letters. Repetition without an arrow indicates that more of the same crop is added in later years of field use. Arrows indicate continuity of perennials and fallow. This note applies also to the other tables of crop sequences below.

FIELD TYPES OF THE TRADITIONAL (IDEAL) CROP SEQUENCE

Based on the dominant crop sown, planted, or remaining, the following field types, arranged in chronological order, may be distinguished:

The maize field (first year field); the manioc field (second year field or first year field on *vípopa*); the plantains and bananas field (later than second year field); fallow fields (later than second year field).

The Maize Field. Every year a new field is cleared. The main part of the field is planted with maize, and smaller areas, generally in segregated blocks with little interplanting, along the access path or in the peripheral areas of the field, are planted on other annuals, and the semi-perennial and perennial crops which comprise the dominant crop associations of ensuing years. In the traditional sequence, beans, both climbing and bush varieties, were intercropped with the maize, but recently the Yukpa have adopted the *wátia* practice of making a separate beanfield. Sometimes, but rarely, manioc, and plantains and bananas are intercropped with the maize.

The Manioc Field. When manioc is planted on *vípopa* soil it is a first year crop on fields which are cultivated for one year only before being returned to fallow. More commonly, manioc comprises the dominant crop grown on second year fields (third year if beans start the sequence). Manioc is commonly grown intercropped in rows with plantains and bananas. Sometimes maize is interplanted in rows in part of the manioc field, but it is never sown in the section in which manioc and plantains and bananas are planted, because it does not yield well under these circumstances. Although gourds and squashes are often scattered between the rows of manioc, annual crops comprise only a very minor fraction of the assemblage present in these fields. The manioc field includes, in addition to the dominant crop, and the concurrently planted crop, a number of perennial crops which were planted during the first year of the field's use: plantains and bananas, sugar cane, pigeon peas, millet, and yautia.

The Plantain and Banana Field. The plantains and bananas which were planted in the first and second year fields gradually assume dominance, so that by third year of use, they comprise almost monocultural stands. By this time too, the area of the field is considerably reduced and at least 30 percent of the original field is reclaimed by *rastrojo.* The *rastrojo* gradually infiltrates the crop which is weeded only casually and irregularly. Sugar cane, which was also planted in part of the first year field, now comprises a small area alongside the plantains and bananas, and small clumps of millet and remnants of manioc and yautia are sometimes found. Normally no further planting takes place in the field because lower growing crops would not thrive in competition with the plantains and bananas.

Fallow Fields. In the normal sequence of field use *rastrojo* is allowed to reclaim part of the field after manioc has ceased to be the dominant crop. After an indefinite number of years, when newer plantain and banana fields are yielding, the old stands are permitted to become swallowed up by the fallow growth. Theoretically within the concept of shifting cultivation, all land not in cultivation, which has not been removed from the cultivation cycle by a specific noncultivation form of land use, should be regarded as fallow. This concept is followed here for the Yukpa territory, although, of course, it is almost impossible to distinguish old field boundaries *per se* in the rapidly growing *rastrojo*.

TABLE 7

Yukpa Crop Sequences: Interrupted Sequence

Season 1	Season 2	Season 3	Season 4	Season 5
MAIZE	FALLOW	FALLOW		
	with	with		
Plantains	Plantains	Plantains	PLANTAINS	PLANTAINS
& →	& →	& →	& →	&
bananas	bananas	bananas	BANANAS	BANANAS
Sugar →	Sugar →	Sugar →	Sugar →	SUGAR
cane	cane	cane	cane	CANE
Beans			Maize	
Squash			Manioc	Manioc
Gourds			Yautia	
Manioc			Squash	Squash
Yautia			Gourds	Gourds
Millet			Millet	Millet
			Plantains	Plantains
			&	&
			bananas	bananas

VARIATIONS FROM THE TRADITIONAL (IDEAL) SEQUENCE

Variants of the ideal sequence and resultant land use type are very frequent and occur as the result of illness, injury, taboo (for example, against asking for seeds and either the need to move from a site on the death of a person or to abstain from labor), social strife, and, above all, recent changes resulting from integration of *wátia* methods into the traditional system.

A large number of possible variations occur, which may be grouped under the following headings: interrupted sequences; truncated sequences; expanded sequences; and diverted sequences.

Interrupted Sequences. This is defined as a normal sequence in a field which is interrupted for one or two years and then resumed or restarted (with the traditional first year crop assemblage) and carried to the normal or slightly modified conclusion. The ideal sequence may be interrupted during its course for a number of reasons, at which time the field enters a fallow phase. The most frequent sociocultural reasons for interruption include the loss of labor through illness, injury or death, and intra-group strife, which forces the cultivator to leave the area to which, only after a period of time, he may return. Interruption results also from ecological causes which necessitate a short period of fallowing to restore desirable properties to the field, and from the failure of cut material to burn properly after several attempts (*vide supra,* p. 77). Only fields which have been fallowed after one year of use and then normally cycled are classified under this heading. Fields fallowed after a two year cultivation sequence are classified as truncated.

An example of an interrupted sequence is given in table 7. In this case the cultivator was forced to flee from the Maracá settlement at Socorpa because he had abandoned his own wife and run off with the wife of another. Although the woman had instigated the affair, as so often occurs among the Yukpa, the man was held responsible. After a period of two years he was permitted to return to the settlement, but, naturally his relationship with many other members of the subtribe is cool.

Truncated Sequence. These are sequences in which the field is fallowed without first completing all of the steps of the ideal field sequence. Truncated sequences are distinguished from those which are interrupted by not being resumed or reinitiated after one or two seasons of fallow. Rather, the truncated field is completely given over to fallow (table 8).

Truncated sequences result from both social and ecological causes. Among the Maracá there are many examples of truncated sequences resulting from attempts to cultivate unsuitable terrain. Landslides, resulting from the cultivation of very steep, unstable slopes are very frequently encountered in Yukpa country. When first year field sites are found to be slide-prone, cultivation is terminated at the end of the first season, and the plot permitted to return to *rastrojo*. If slide-prone areas are recognized in advance they are not cleared for cultivation but rather are left as game preserves.

Fields are also fallowed if the first year of cultivation yielded only a meager harvest, the reasons for which could be traced to excessive weed infestation and often, soil wash-out. Fields bounded by savanna and fern-dominated fallow associations frequently undergo a truncated sequence.

Often, however, in such cases, a small portion of the field furthest from the source of infestation is used for a second season before it also becomes too infested with noxious weeds to be usable.

The other two field-use sequences encountered among the Yukpa have been developed to accommodate innovations in the subsistence system adopted as a result of contact with *wátia.*

TABLE 8

YUKPA CROP SEQUENCES: TRUNCATED SEQUENCES

Season 1		Season 2	Season 3	Season 4	Season 5
			FIELD A		
PLANTAINS	L	PLANTAINS	FALLOW →	FALLOW →	FALLOW
& →	A	&			
BANANAS	N	BANANAS			
	D				
	S	Fallow			
	L				
	I				
Manioc	D				
Yautia	E				
			FIELD B		
MAIZE	E	YAUTIA	FALLOW →	FALLOW →	FALLOW
	X				
Beans	C	Maize			
	E				
Squash	S W				
	S E				
Gourds	I E				
	V D				
Manioc	E S				
Yautia					

The Expanded Sequence. Expanded sequences involve the incorporation of newly acquired beans and pasture grasses; and the modification of the ideal sequence to permit two harvests of established crops during one annual cultivation cycle.

The incorporation of newly acquired beans represents the commonest type of expanded field-use sequence. Separate fields are cleared for planting bush beans. As a consequence the traditional climbing varieties are rapidly decreasing in importance. Beans have been integrated into the field-use sequence by sowing them as a dominant first year crop, together with smaller areas planted in clumps of sugar cane and some patches of plantains and bananas. Maize, with its associated small areas of annual and semi-perennials, then enters the sequence as the dominant crop in the second year. The adoption of *wátia* bush bean varieties has led to an extension of the length of field cropping by one year.

TABLE 9

YUKPA CROP SEQUENCES: EXPANDED SEQUENCE

Season 1	Season 2	Season 3	Season 4	Season 5
		FIELD A		
BEANS	MAIZE	Part fallow →	Fallow →	Fallow
Yautia	Yautia			
Plantains & bananas →	Plantains & bananas →	PLANTAINS & BANANAS →	PLANTAINS & BANANAS →	PLANTAINS & BANANAS
Sugar cane →	Sugar cane →	SUGAR CANE →	SUGAR CANE →	SUGAR CANE
	Manioc			
	Plantains & bananas →	PLANTAINS & BANANAS →	PLANTAINS & BANANAS →	PLANTAINS & BANANAS
		FIELD B		
MAIZE				
Beans				
Squash				
Gourds	Maize			
Manioc →	MANIOC	FALLOW →	FALLOW →	FALLOW
Yautia →	Yautia			

Many of the Yukpa sow part of the field in pasture grasses in the last year of the normal sequence of field use. This is done on a very small scale in most cases to provide forage for the one or two beasts of burden or cows that they maintain. The forage provided is more nutritious than that obtained from savanna grasses, many of which the animals find unpalatable. These sown pastures in the majority of instances are not maintained and within a few years are heavily infiltrated by fallow growth.

Still more recent than modification induced by the introduction of new bean and pasture grass species has been the expansion of the ideal field use sequence in the *t. caliente* of the lower Tukuku and Negro valleys and the Valley of the Motilones, to incorporate a second maize crop during the second year of the field sequence. During the rainy season, sections are cleared in the second year manioc field where plantains and bananas have not been planted and sown with maize. One crop is taken and the field permitted to return to fallow.

TABLE 10

YUKPA CROP SEQUENCES: DIVERTED SEQUENCES

Season 1	Season 2	Season 3	Season 4	Season 5
		FIELD A		
MAIZE	Coffee →	COFFEE →	COFFEE →	COFFEE
Manioc →	MANIOC			
Yautia →	Yautia			
Beans				
Squash				
Gourds				
		FIELD B		
MAIZE				
Manioc →	MANIOC →	COFFEE →	COFFEE →	COFFEE
Yautia	Yautia			
Plantains	Plantains	PLANTAINS	PLANTAINS	PLANTAINS
& →	& →	& →	& →	&
bananas	bananas	BANANAS	BANANAS	BANANAS
Beans	Plantains			
Squash	&			
Gourds	bananas			
		FIELD C		
MAIZE				
Manioc				
Pasture →	PASTURE →	PASTURE →	PASTURE →	PASTURE

Diverted Sequences. Diverted sequences, unlike the expanded variety described above, involve radical changes in the cultivation system. In a few cases in the Valley of the Motilones and the lower Negro Valley, coffee bushes and pasture are planted at some stage during the normal sequence, and fallow vegetation is not permitted to return. In some instances coffee plantations and pastures are planted independently of the fields worked by shifting cultivation, and the two systems are operated jointly and mutually exclusively by the same cultivator.

In the majority of cases where coffee bushes are planted, not all of the vegetation is removed when a field site is cleared. Despite the intention to plant a normal first year maize-dominated crop assemblage, sufficient foliage is left to provide shade for the coffee to be planted in the second year. In some instances the normal sequence is followed to the end of the second or third year and then coffee is planted beneath the shade of plantains and bananas.

On some holdings pasture grass is planted in the first year of cultivation, together with corn and manioc, whereas plantains and bananas and pineapple are planted in separate fields. In the extreme example of this type of sequence the system employed may no longer be regarded as shifting cultivation. The cultivator is now specialized in cattle and cash crop raising, with subsistence crops grown on first year fields purely as a sideline. This example is discussed at length below, pp. 149–153.

Field-Shift Patterns

There is no single adjective which may be employed to describe succinctly Yukpa field-shift patterns. The random,[1] unidirectional, and circular field-shift patterns summarized by Spencer (1966:144–147) are all to be found in Yukpa country, but not all of them enter into the case studies discussed later in this chapter.

FACTORS NECESSITATING FIELD-SHIFT

In the Yukpa system of shifting cultivation, fields are shifted primarily in order that fallow vegetation may shade out noxious weeds which have invaded cultivated fields and reduced crop yields, and also so that soil fertility may be restored. The other major ecological reason for field shift is to reduce damage by crop plant pests, which sometimes constitute especially grave problems in the second year of field use. There are also a number of sociocultural factors which govern the shift pattern of fields.

For the Yukpa two factors are paramount among the selection criteria for new fields: The land tenure system; and the factors of site ecology.

The Land Tenure System. The pattern of land ownership is the basic factor governing the distribution of field sites. A cultivator may plant only in his own land, in land borrowed for a specific number of seasons, or in unclaimed land sanctioned for his use by the rest of the subtribe. In the

[1] Unfortunately this term somewhat misrepresents the situation, for it is to be doubted if the selection criteria and the subsequent distribution of fields are ever "random." Spencer (1966:145) described, in a somewhat contradictory manner, the random shift, "in which succeeding fields are picked by abstruse criteria consisting chiefly of omens and signs." For some groups of people omens and signs are often important and specific site selection criteria, and fields distributed in accordance with them, cannot truly be said to be randomly distributed.

normal course of events clearance of unclaimed land is exceptional, even if land owned or borrowed is at a considerable distance from the pivotal settlement.[2] Cultivators in many instances, though I have not documented any specific outstanding examples, may spend a considerable number of hours travelling between their homestead and their field, although in such cases, to avoid an undue loss of time, a secondary, temporary field house is generally constructed. Most instances of fields located at long distances from the settlement result from inheritance. Such fields have to be worked when the local biophysical conditions are suitable, because land closer to the settlement is, in most cases, owned by other cultivators, or perhaps has been rendered unsuitable for immediate needs by recent cultivation or escaped burns.

The distant location of the fields with respect to the homestead has occasioned considerable comment from various authors (for example, de Booy, 1918*b*:395). In general, it may be safely said that past commentators failed to appreciate the operation of both the cultivation and land tenure systems, preferring instead to attribute the situation to the endemic warfare beween the subtribes and the need, for defensive reasons, to locate the fields and the homesteads far apart.

The Factors of Site Ecology. Factors governing field site ecology and the requirements of the planted crops have already been examined in detail in the preceding chapter. Suffice it to say here that within the limits of the land owned and otherwise obtained, field sites are selected above all and with specific reference to: The nature of the fallow vegetation occupying the land, and the surrounding plant associations; the requirements of the crops with respect to topographic, edaphic, and "aspect" factors; and the relation of the proposed sites to the overall sequence of field crop shifts.

Basically it may be said that the factors which lead initially to the selection of first year field sites also operate to cause the fragmentation of them when they evolve into the smaller second and third year fields. Most Yukpa cultivators are unable to locate their fields in a contiguous manner; and, as a result, the dominant land use pattern is one of fragmented and scattered fields. In many instances, however, the distance between cultivated fields of various ages is not great.

[2] For the purposes of this discussion the "pivotal settlement" is defined as the permanent homestead in which a cultivator dwells and about which the fields "turn."

The Dooryard Garden

Dooryard gardens among the Yukpa are highly variable in importance, in the number and character of the plant species contained, and in the degree of separation from the field.[3] Some homesteads are surrounded by only rudimentary and casually cared for gardens with only one or two species growing either against the wall of the house or within the fallow surrounding the house site. Other gardens comprise a large number of plant species which occupy a fairly large area around the house site and are segregated from the surrounding fallow and field vegetation by definite boundaries. Some dooryard gardens contain only plants which have ritual and technological usage together with longer-lived perennial species, whereas others contain, in addition, small quantities of the dietary staples. Some gardens are truly "dooryard" whereas others are located along the axial path in the main field, and beside stands of plantains and bananas which are often maintained in cultivation for many years after the rest of the field has returned to fallow. Generally, the Yukpa garden is located near the house because this ensures a greater degree of permanence and accessibility than does a garden located in a field.

GARDEN PLANT SPECIES

The species planted by the Yukpa in the dooryard garden may be grouped in the following manner: Traditionally sown annuals; traditional perennial species; recently introduced species; and plants generally grown in the shifting field.

Traditionally Sown Annuals. Annual plants traditionally sown in the dooryard garden are tobacco, gourds, squash, and balloon vine (pl. 14). Occasionally climbing beans are sown to grow up specially constructed trellises (pl. 10). It is important to note that none of these plants is grown exclusively in the dooryard garden. Indeed, the larger percentage of these crops is derived from the field.

Traditional Perennial Species. As would be expected from the greater degree of permanence afforded by the homestead site, the majority of plants

---

[3] The range of garden types among the Yukpa is illustrated by maps 3, 4, and 5. Map 3 illustrates the basic garden devoid of any staple crop plants; map 4 shows the increase in complexity of the garden with the introduction of field crops and plants obtained from the *wátia;* and map 5 illustrates the newest type of garden, that found in the *Strassendorf* west of the mission at El Tukuku.

traditionally grown in the dooryard garden are perennials. However, a few perennials are also planted in accessible parts of the field, particularly the red pepper, which should be conveniently handy at all times both in the house and field for driving away evil spirits. The traditional perennial plants most commonly grown in the dooryard garden are: castor bean, annatto, hibiscus, red pepper, calabash tree (in the *t. caliente*), pigeon peas, and the agave.

Recently Introduced Species. New varieties of traditional staple crops obtained from contact with the *wátia* are first planted experimentally in a limited area of dooryard garden to see how they taste and if they yield well. If they are liked, and the input of labor necessary for their cultivation is deemed worthwhile, they are later planted in a larger field.

In addition, recently introduced perennial fruit trees are planted in the dooryard garden for the same reasons as are the traditional perennials. A few coffee bushes are sometimes planted near the house, and a small number of pineapples are grown by some Irapa and Rionegrinos in the dooryard garden, but one Irapa has over 200 plants which he is raising as a cash crop. Tomatoes (which are occasionally also grown in the field) and decorative garden plants, such as "stocks," are occasionally found in the garden.

Field Plants. Very frequently, field crop plants which constitute the main dietary staples of the Yukpa are also grown for convenience, or to provide a small early harvest before the main harvest, in or near the dooryard garden. Plantains and bananas are the field crops most commonly encountered in the garden, where they are also valued for their shade. Barbasco is also planted in the shadier parts of both the garden and the field. Clumps of millet are planted in the garden close to the house. In this manner birds, which are attracted to its ripe seeds, can be conveniently shot from the shelter of the house, which doubles as a blind.

THE USE OF GARDEN PLANTS

It becomes somewhat difficult to distinguish between a true dooryard garden and a field which is merely adjacent to the homestead.[4] In some societies the nature of the treatment given to each may provide a basis for separation, but among the Yukpa the criterion for the distinction between

[4] There is no basis for calling this an "in-field," because it is in no way treated distinctly from any other field.

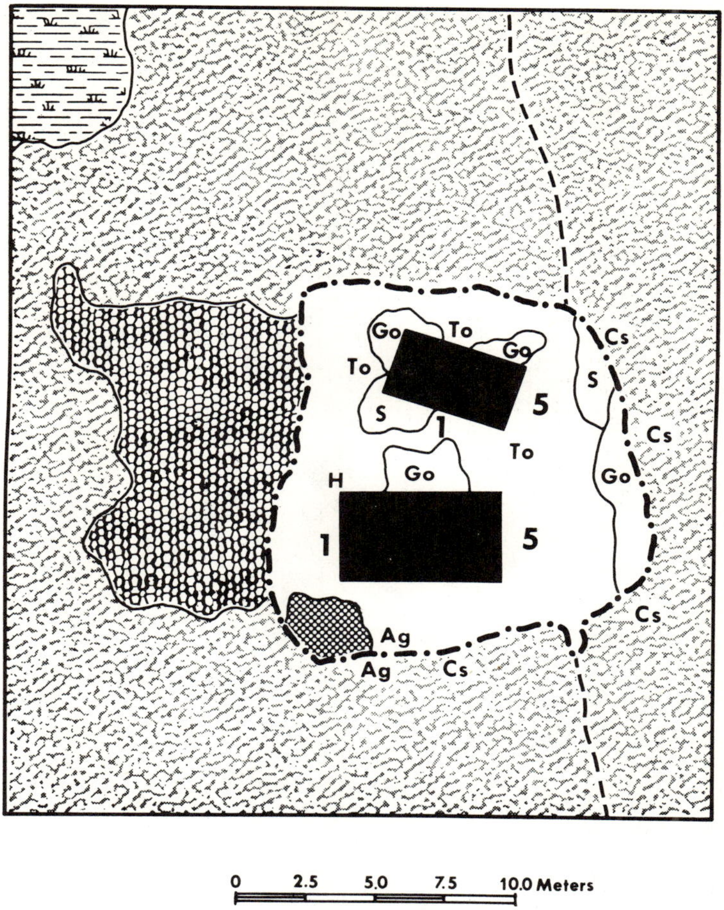

Map 3. Maracá dooryard garden.

dooryard garden and close-by field is the character of the majority of species planted. Thus, although some field crops may be said to be growing in a dooryard garden situation, the majority of those planted near the homestead should still be regarded as field crops.

In the Yukpa autosubsistence system where primary economic support is derived from the cultivated fields, the role of the dooryard garden is somewhat specialized. It is the main place for growing plants with ritual usage (red pepper, datura and hibiscus); cosmetic and ornamental usage (annatto and canna); plants with important technological usage (castor

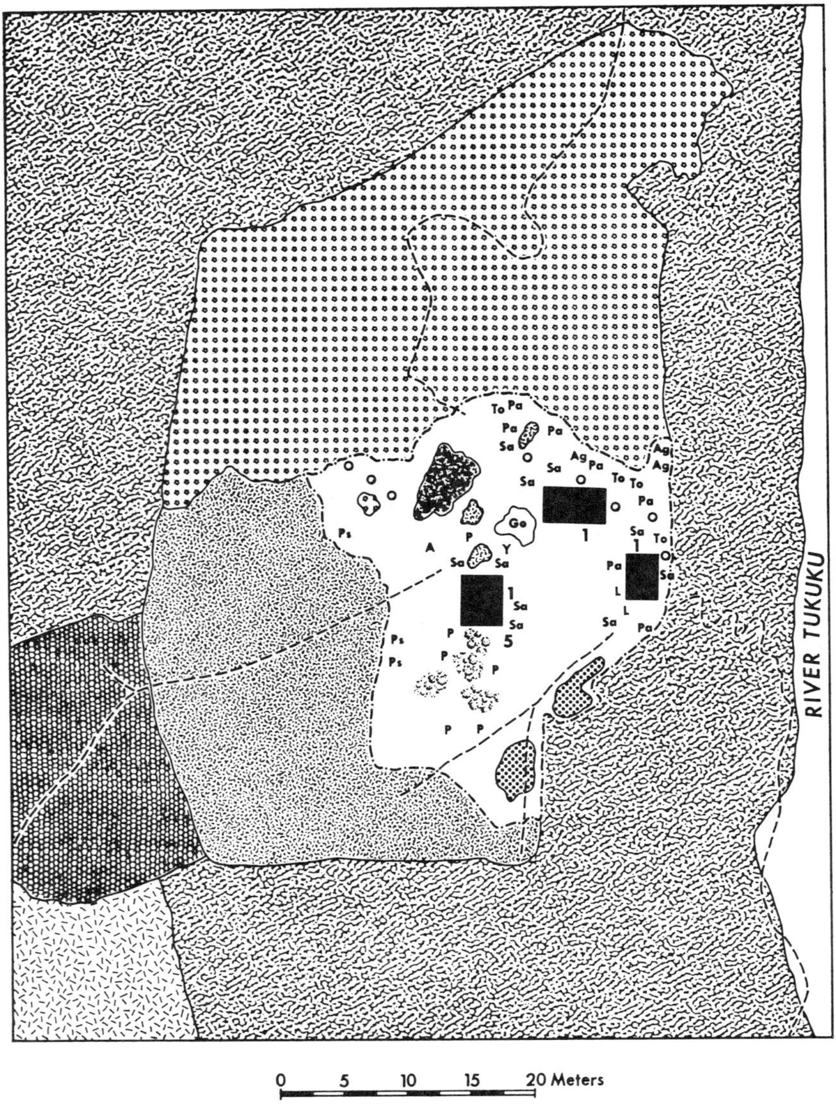

Map 4. Irapa dooryard garden.

bean, barbasco, agave, calabash tree, gourds, millet); and plants, such as fruit trees, some of which, besides requiring a long growing period before reaching maturity, are valued for the shade which they provide.

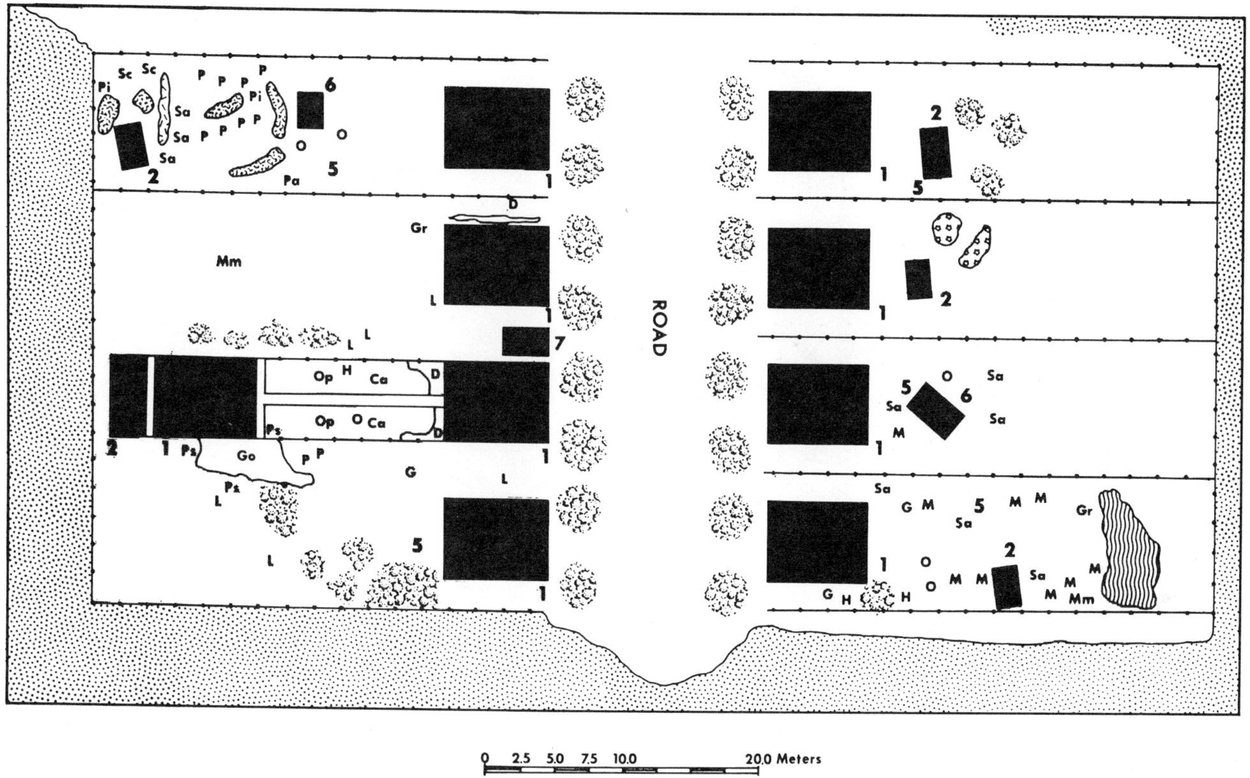

Map 5. Dooryard gardens at El Tukuku mission.

THE ROLE OF THE GARDEN

Besides providing a specialized area for the growth of particular species, the dooryard garden is important for two reasons. First, it provides a spot where the growth of newly acquired plant species and varieties can be observed, and the seeds and clones of particularly good specimens selected if the plant is judged suitable for incorporation into the field crop assemblage. Second, the dooryard garden is useful in that certain plants may be grown conveniently close to the homestead, obviating the need to walk to the field which may be far away. This is particularly important in the case of red pepper.

LABOR AND CULTIVATION TECHNIQUES

In terms of labor and cultivation techniques employed, the dooryard garden differs little from the cultivated field. The woman does, however, undertake a marginally greater percentage of the work load in the garden, and has more contact with the garden plants because of the longer periods of time that she spends around the homestead.[5]

The garden plants receive no special treatment. They are cared for in exactly the same manner as are the field crops. It should be noted, however, that the dooryard garden plants do benefit from the fertilizing effects of household refuse, which is casually tossed around the perimeter of the house site.[6] Field crops, on the other hand, do not receive any form of manure or fertilizer, other than the ash left by burning.

Land Use Case Studies

The elements of Yukpa land use have been described. In the remainder of this chapter, I will endeavor to draw these elements together to show the overall patterns of the cultivation landscape, using five condensed, comparative case studies.

THE GARDEN FAMILY OF AMÚŠMO (MARACÁ)

This "garden family" of which Amúšmo is the head, was chosen for detailed presentation as an example of an average Yukpa fundamental economic unit (nuclear family) which depends primarily on the products of the cultivated field for its livelihood. Other autosubsistence activities

[5] The cultural and historical implications of the woman-dooryard garden plant relationship have been discussed by Sauer (1952).

[6] This is not done with the motive of benefiting the garden plants, but merely to dispose of refuse.

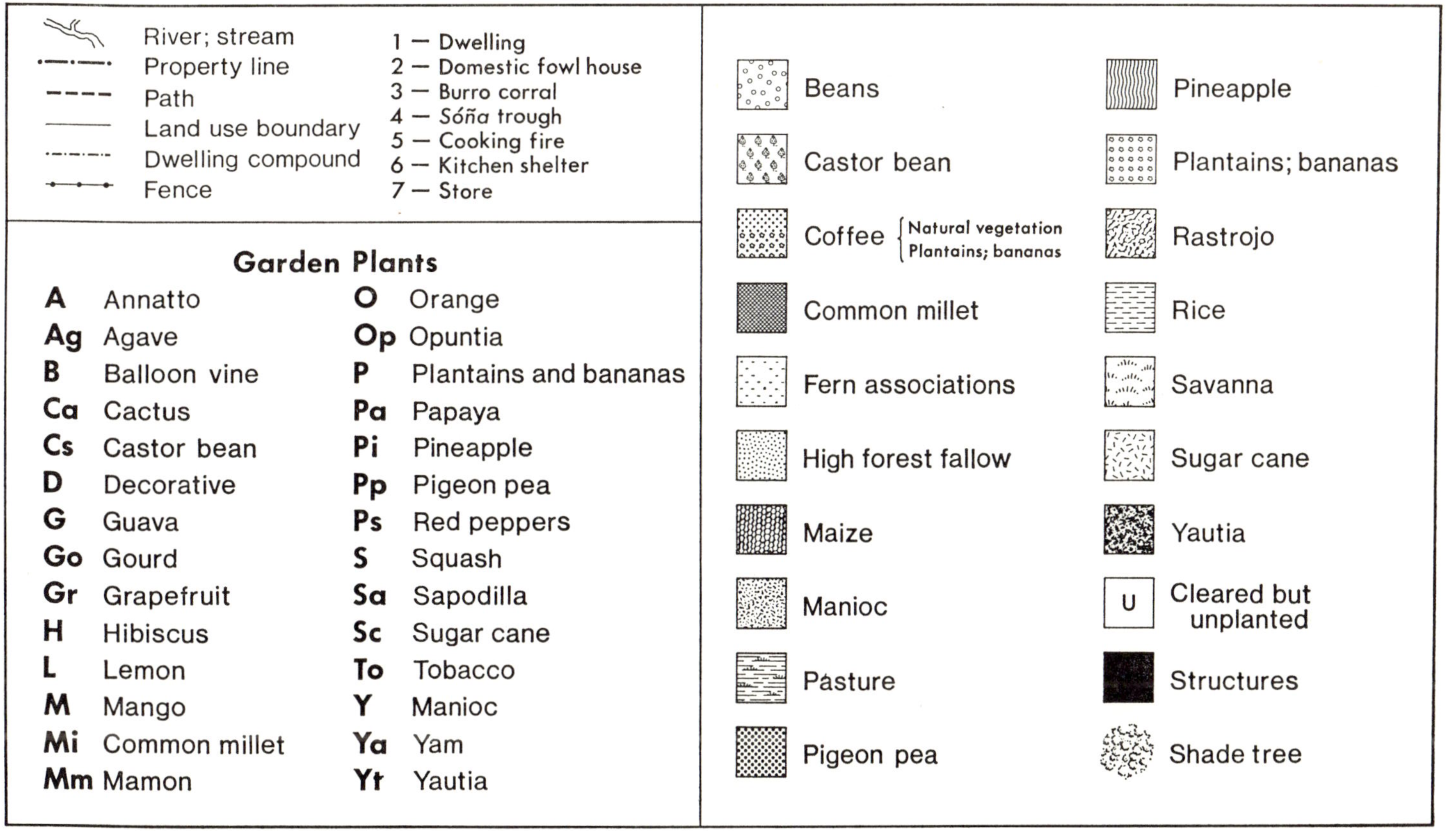

Legend for dooryard garden and land use maps.

play a secondary, but nevertheless, important role. It was also chosen for in-depth study because of the reliability of the economic statistics which could be obtained. The head of the household occasionally works with the other adults of the family on the surrounding *fincas* of the *wátia,* especially during the bean harvest. Occasionally too, he sells hats and baskets to the missionaries. In most ways this is a traditional household which shows neither incorporation of *wátia* subsistence elements nor the abandonment of Yukpa ways to subsist from mission handouts.

The garden family comprises a work group of four, plus an old maternal aunt[7] of Amúšmo, his two small children and one grandchild. In addition, two other adults, who are members of another garden family which lives only half an hour away, occasionally spend several days visiting this house and eating there. Amúšmo is a man of some influence among the Maracá; thus he is frequently feeding visitors.

The labor force of the garden family is of the traditional Yukpa type. It comprises Amúšmo, the head of the household, and his son-in-law, who together undertake the bulk of the cultivation labor, aided by Amúšmo's wife and daughter, who help with the sowing and undertake all of the maize harvest.

The total cultivated area producing food crops comprises 4.047 ha.,[8] divided into the maize field, the manioc field, and the plantain and banana field.

The Maize Field. The field was cut in tall forest fallow on a gently sloping river terrace, with *sásare* soils. Weed infestation, which normally is not serious in such clearings, is in this case exacerbated because the upper part of the field borders on savanna and is heavily infested with grasses.

Of the total first year field, 2.971 ha. was sown with maize, and 0.321 ha. planted in other crops, maize and other crops being interplanted on 0.130 ha.

[7] She died in November 1969 after I had left the area (Personal communication: P. Hermenegildo Pérez, December, 1969).

[8] This gives a subsistence area of 0.526 ha. per capita of the garden family. This is slightly higher than the Yukpa average (*vide infra*).

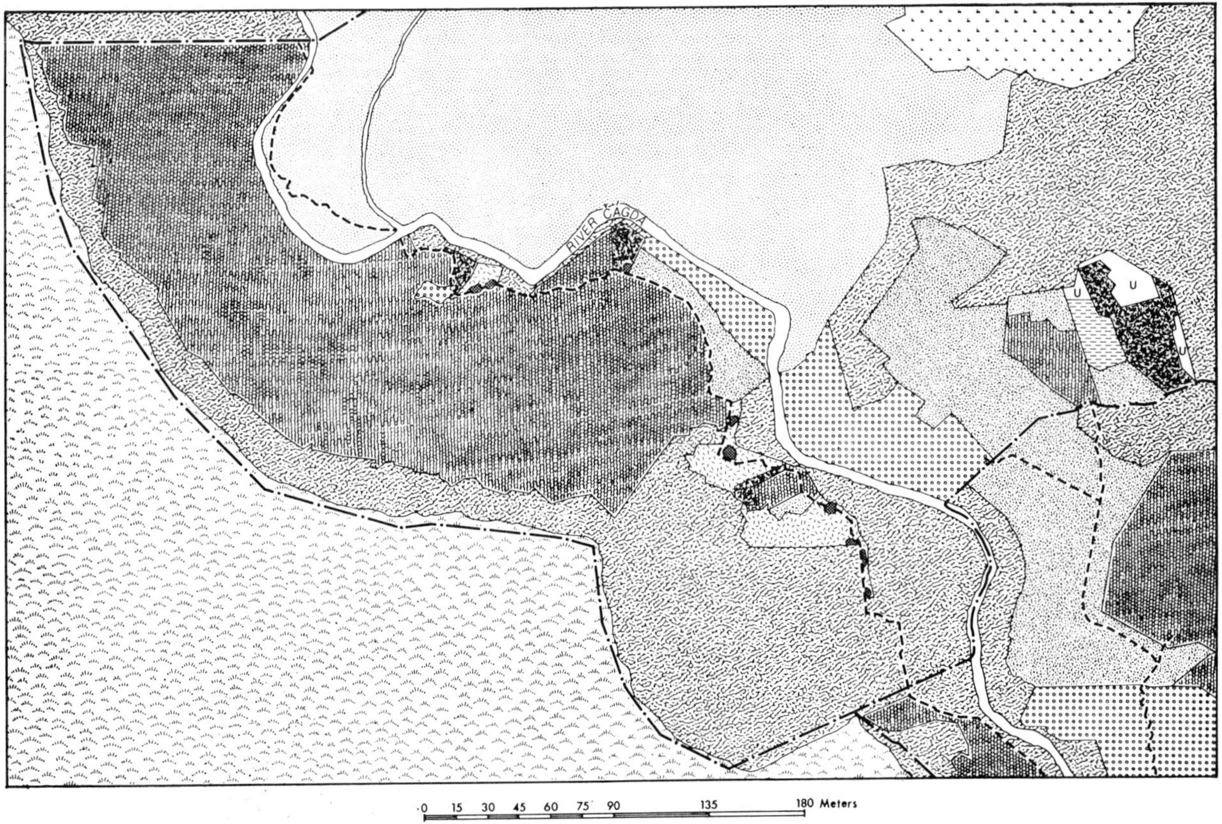

Map 7. Land use: Amúšmo—1969.

The planting pattern (map 7) comprises:

Maize	2.841 ha.
Maize and manioc	0.025 ha.
Maize, manioc, squash and gourds	0.093 ha.
Maize, plantains and bananas, and yautia	0.012 ha.
Manioc	0.010 ha.
Manioc and yautia	0.021 ha.
Plantains and bananas, and yautia	0.001 ha.
Sugar cane	0.097 ha.
Sugar cane and millet	0.016 ha.
Sugar cane, yautia, and millet	0.013 ha.
Sugar cane and yautia	0.013 ha.
Sugar cane, manioc, yautia, and millet	0.007 ha.
Yautia	0.013 ha.
Total	3.162 ha.

The non-grain crops in the first year field are planted in the most accessible locations along the axial path. Yautia and manioc are also planted apart from other crops because their heavy foliage puts other crops at a disadvantage. Tobacco, yams, sweet potatoes, and the occasional canna are scattered along the axial path, because this cultivator regards them as only minor crops. During the second year of use the upper portions of the field were to be fallowed to check the incursion of noxious grasses.

The Manioc Field. The field comprises 0.444 ha. and constitutes the remainder of the previous year's maize field, some of which was permitted to return to fallow, part planted in manioc, and part (0.390 ha.) loaned to a nephew (*vide infra,* p. 148) for manioc and experimental rice cultivation. The manioc field in this instance has not yet been interplanted with plantains and bananas.

The Plantain and Banana Field(s). The plantain and banana fields are fragmented. A patch (0.194 ha.) opposite the house was planted in 1966, and some (0.328 ha.) of the 1967 field was planted in plantains and bananas during the first year of cultivation and added to during the second year. Although part of the 1967 field was permitted to return to fallow in 1968, a section was cleared again in 1969 and 0.076 ha. was planted in plantains and bananas. The total fragmented plantain and banana area, with the exception of that in the first year field, comprises 0.598 ha.

The total area under cultivation in 1969 thus comprised:

Maize	2.841 ha.[9]
Manioc	0.454 ha.
Plantains and bananas	0.598 ha.
Sugar cane	0.097 ha.
Yautia	0.013 ha.
Combined plants	0.201 ha.
Total	4.204 ha.

Interplanting and Crop Segregation. In the Yukpa cultivation system interplanting is mainly restricted to plantains and bananas with manioc in the second year of field use, and occasional interplanting of maize and manioc in both first and second year fields. Small-scale interplanted crop combinations are plantains and bananas and yautia, sugar cane and yautia, manioc and yautia, sugar cane and millet; and, in the patches (but not all interplanted in the same patch) and scattered in the maize field, squash, gourds, millet and tobacco. In this case study only 0.205 ha. of the total area in cultivation consists of polycultural plantings. The remainder is planted in monocultural blocks of crops, within which the plants are further separated by varieties. The staple food crops are always segregated by varieties.

ČAKÓČA AND WIKÓČUMO (MARACÁ)

This case illustrates the commonest type of cooperative unit in use among the Yukpa, the communal cultivation by two brothers. The "title" to land is held independently by the two, but the working of the land and rights to the production is shared equally. To preserve the equality of land value, fields are not cut in one brother's property two years in succession. Traditional cultivation and land use techniques have been retained and the adoption of *wátia* ways is minimal. One of the brothers, however, spends time working in the mission fields, where his presence is virtually continuous at harvest time.

The combined cultivated area (5.666 ha.)[10] of food crops is large by Yukpa standards, but the combined garden family is ten people and the cultivated area per capita is 0.542 ha. The garden family comprises three

[9] Area for maize, manioc, plantains and bananas, sugar cane, and yautia represents monocultural area only.

[10] Owing to rounding, the total cultivated area does not exactly equal the sum of the individual parts mentioned in the following pages.

adult males (two brothers together with the son of one of them), who undertake the cultivation labor, two wives and five small children.

The land use characteristics of the joint holding are somewhat atypical for the Yukpa, in that combined working permits a more judicious selection of fields to suit particular crop characteristics. Thus, the cultivated plots are somewhat more fragmented within the holding, particularly in second year fields, than is the general rule (see map 8).

The Maize Fields. This garden family had two maize fields newly cleared in 1969. A small field of 0.051 ha. was cleared near the house site to give an early maize crop, before the main field was producing. This was needed because the previous year's store of maize had all been consumed. Gourds were interplanted in parts of this small field of early corn, and also along the perimeter of the house site. The main maize field, comprising 3.169 ha., was sown exclusively with maize.

Second Year Fields. The second year fields focus on field houses, built for crop storage and for temporary living quarters during the arduous phases of field clearance. Because of the greater amount of land available to them as a result of cooperative working, the cultivators were able to select only the portions of the 1968 field with the richest soils for second year use. Thus, the second year field was divided into six small, carefully selected fields surrounded by an area which was permitted to return to fallow growth. The soils of the second year field were sufficiently rich in places to make it worthwhile to selectively replant in maize.

The second year fields, which comprise 0.408 ha., are divided as follows:

Plot 1 (0.151 ha.):	Beans	0.035 ha.
	Maize	0.041 ha.
	Manioc	0.075 ha.
Plot 2 (0.123 ha.):	Maize	0.090 ha.
	Maize and manioc	0.014 ha.
	Millet*	0.002 ha.
	Yautia	0.019 ha.
Plot 3 (0.069 ha.):	Manioc	0.009 ha.
	Yautia	0.060 ha.
Plot 4 (0.065 ha.):	Maize and manioc	0.065 ha.
Plots 5 and 6 (0.007 ha.):	Millet*	0.002 ha.
		0.005 ha.

*Millet is totaled separately because it is not a food crop. (Total millet area = 0.009 ha.)

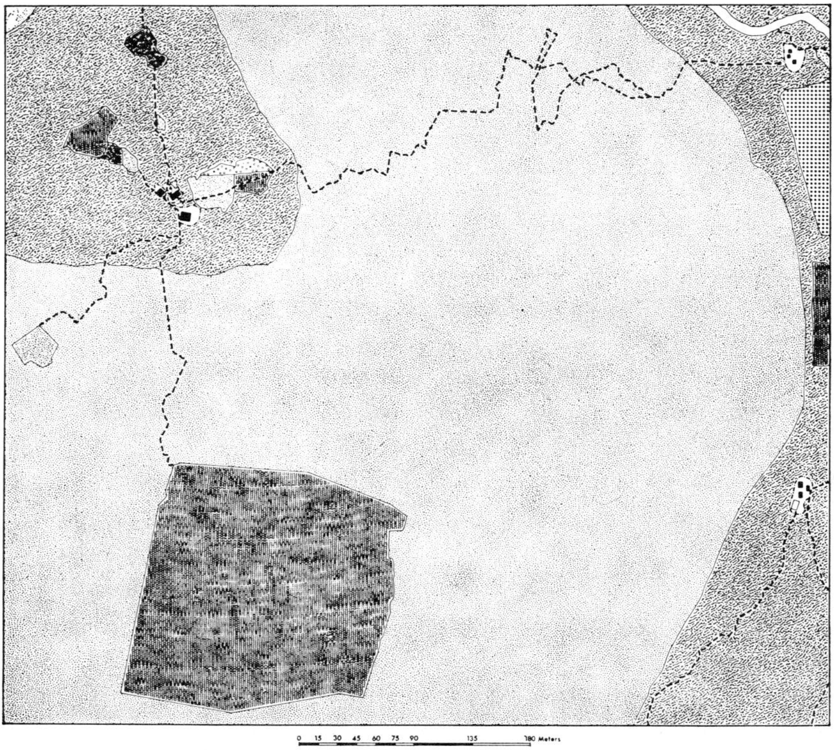

Map 8. Land use: Čakóča and Wikóčumo—1969.

Interplanting is minimal; only maize and manioc (plots 2 and 4) are thus cultivated.

Plantain and Banana Fields. The single plantain and banana field of 2.048 ha. is located in the shelter of the valley bottom. It is composed of plants which range over three years in age. Initially some suckers were planted when the field was in maize during the first year of use, and more were added during the second year. Finally, during the third year, the field was extended and yet more plants were added.

THE GARDEN FAMILY OF CAMÁŠAŠ (MARACÁ)

The garden family was chosen because it exhibits an extremely low per capita area of 0.102 ha. The below average cultivation production is compensated for by an increased emphasis on hunting and gathering, and a dependence on the mission and other families for handouts of food. Furthermore, this family is illustrative of one aspect of the collapse of the Yukpa social system, where obligatory labor rules applied to kinfolk are ignored. Although this garden family shares a residence site (pl. 15) with the old woman's brother, the brother's garden family is unable to contribute much to their sustenance.

The products of a total cultivated area of 0.922 ha. are used to feed a garden family of nine, composed of a very old man, who is unable to work, and his aged wife, together with his daughter and son-in-law (who undertakes all of the cultivation labor) and their three small children. In addition, two young males, the nephew of the old women and the unmarried son of the old man, live in the house but contribute very little to the subsistence production.

The two young males who form part of Camá̌šǎs's family refuse to help him with the cultivation labor as the result of a family squabble,[11] with which I am not entirely familiar. They spend much time hunting, but it is thought by the rest of the garden family that they cook and eat the products in the forest to avoid sharing them with the other family members. Yet, however antisocial this habit may be, it does lessen somewhat the demands made on the field products, as do the occasional meals provided by the missionaries to the two old people, who occasionally weed the nuns' kitchen garden at the Socorpa mission station. Camá̌šǎs works two small fields (map 9), both of which are devoted mainly to maize production. (For this reason they are denoted by age and not in terms of the dominant crop.)

The First Year Field. The first year field, which comprises 0.522 ha., is entirely interplanted with beans, maize, manioc, and squash. The beans, a bush variety, were planted first. When they were flowering maize was interplanted. After the beans were harvested manioc and squash were interplanted with the maize.

The Second Year Field. Part (0.024 ha.) of this field was planted in plantains and bananas and manioc during the first year; during the 1970 season these will be extended to cover the entire area. The soil (*nóno ķoráčašķ*) was rich enough that 0.375 ha. of this field was replanted to maize in 1969, with squash interplanted over much of the area. There was no third year field in 1969.

[11] It should be noted, however, that although the unmarried son of the old man is obliged to help with the cultivation labor according to Yukpa social rules, the nephew of the old woman is not.

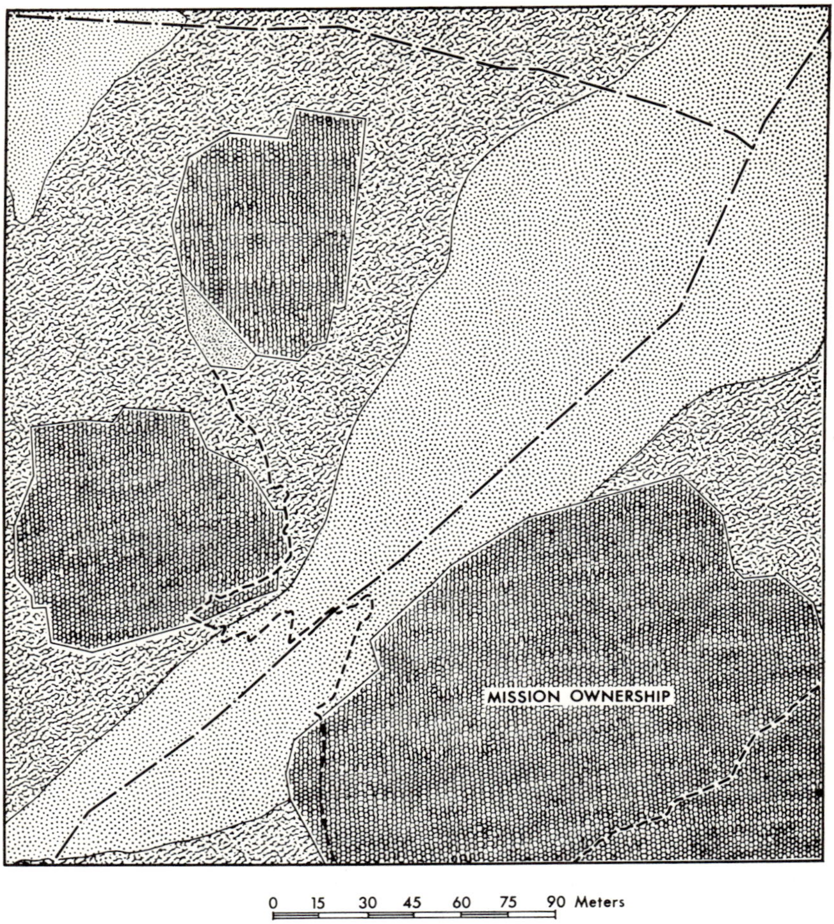

Map 9. Land use: Camášaš—1969.

THE GARDEN FAMILY OF ČÁČO (MARACÁ)

This case study has been chosen to illustrate incipient cash cropping, and the first tentative steps taken by a Yukpa cultivator to explore the market economy of the Colombian *campesino,* and the hiring of labor, together with the traditional borrowing and gift of land. Basically, however, the efforts of the cultivator are directed toward subsistence production, and unlike the next case illustrated, cash cropping is still only a tentative endeavor.

The total area cultivated in 1969 comprised 4.00 ha., of which 2.428 ha. was devoted to subsistence crops. The rest of the area was used as follows: 0.890 ha. produced 907 kilos of maize for sale, and 180 kilos retained as seed corn; 0.615 ha. of beans produced 227 kilos for repayment of a bean seed loan made by the Capuchin priest, 237 kilos for sale, and 90 kilos for retention as seed; an unsuccessful experimental rice crop was planted in 0.046 ha. of borrowed land; and 0.022 ha. was sown in *piríñi* to produce arrow shafts.

Thus, the total area available for the direct subsistence of the garden family of seven, yields a per capita area of 0.347 ha. This low area is compensated for by canned goods and other food items bought from the mission with the money obtained from crop sales.

The garden family, which is composed of the three adults (the cultivator, his wife, and an aged aunt) plus four young girls, is partly supported by the mission kitchen, which provides meals to three of the girls who attend school daily. Thus, this cultivator is able to take advantage of the effective reduction in the size of his garden family and can sow some of this subsistence area with cash crops, without reducing his family's food supply.

The land use pattern of this garden family has been further complicated (map 10) by interrupted field use.

The Maize and Bean Field. This field, which is divided between the two dominant crops, was cleared in February of 1969 in high forest fallow. Its soils are principally sandy *sásare,* on gently to steeply sloping valley sides, with patches of *nóno koráčašk* along the damper and flatter stream margins. On its upper margin it is bordered by savanna and trees killed by an escaped burn when the field was cleared.

The land use of the field, in which there was little interplanting, was as follows:

Beans	0.773 ha.
Maize	1.113 ha.
Plantains and bananas	0.004 ha.
Plantains and taro	0.017 ha.
Sugar cane	0.013 ha.
Yautia	0.015 ha.
Total	1.935 ha.

In the second year (1970) of field use the cultivator planned to cultivate the same crops, but to rotate the maize and beans, and to extend the areas of the associated perennial crops. Because of extreme susceptibility to soil

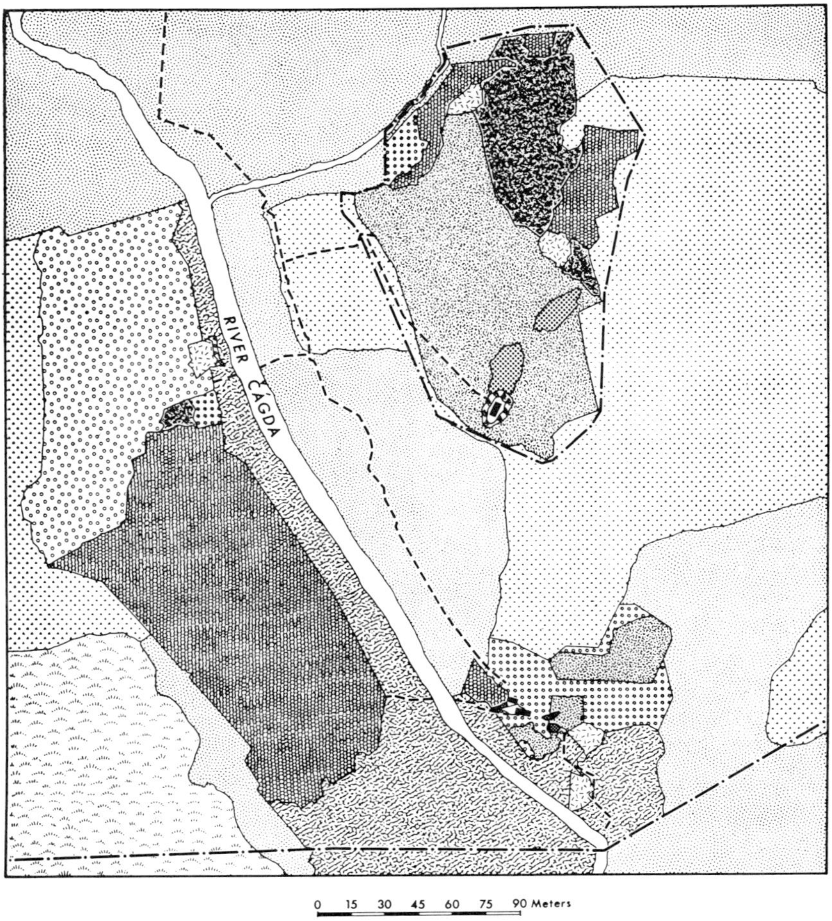

Map 10. Land use: Čáčo—1969.

erosion, however, the field was abandoned to *rastrojo* and he moved to an adjacent valley.

The Manioc Field. Čáčo was chased out of the Maracá settlement for wife stealing, and was absent during 1967 and 1968. There was, therefore, no second year field in which manioc could be planted. Further, despite the help of one hired man, he was not able to clear a special field for manioc. Thus he fell back on the kinship obligations and asked two uncles for the loan of sufficient second year land (which would have been returned to fallow) for the cultivation of manioc and other crops. This borrowed land, which comprised 0.778 ha. (both uncles lending 0.389 ha.) was planted in the following manner:

Maize and manioc	0.102 ha.
Manioc	0.506 ha.
Rice	0.046 ha.
Yautia	0.124 ha.
Total	0.778 ha.

The Plantain and Banana Field. One plantain and banana field located alongside the homestead was the maize field in 1966, and at that time had a small area planted in plantains and bananas (0.988 ha.) and sugar cane (0.271 ha.). During the cultivator's absence *rastrojo* returned. In 1969 the field was cleared again and planted with the mixture of crops:

Maize and manioc and squash	0.016 ha.
Maize and plantains and bananas	0.016 ha.
Plantains and bananas and gourds	0.012 ha.
Plantains and bananas and manioc	0.026 ha.
Total	0.070 ha

The production of plantains and bananas is derived mainly from a separate field of 0.278 ha. on land owned by the cultivator, but located at some distance from the main property. This field, together with a patch of *piríñi* (0.022 ha.), was planted in 1966.

EWA (IRAPA)

The last of these case studies deals with the newest type of land use system among the Yukpa. In the lower Tukuku and Negro valleys a number of Yukpa are gradually moving away from the traditional system of shifting cultivation, as a result of the adoption of permanent field crops and livestock.

The germ of permanent field cultivation is contained within the traditional Yukpa system, in which fields of plantains and bananas are maintained in production for many years. Furthermore, the frequency of house shifting was not closely related to field shifting. Thus, although the introduction of other crops necessitating permanent maintenance requires major adaptive changes in the traditional life style, it is not necessarily inimical to the time-honored system, and in many cases these changes can be made without undue disruption.

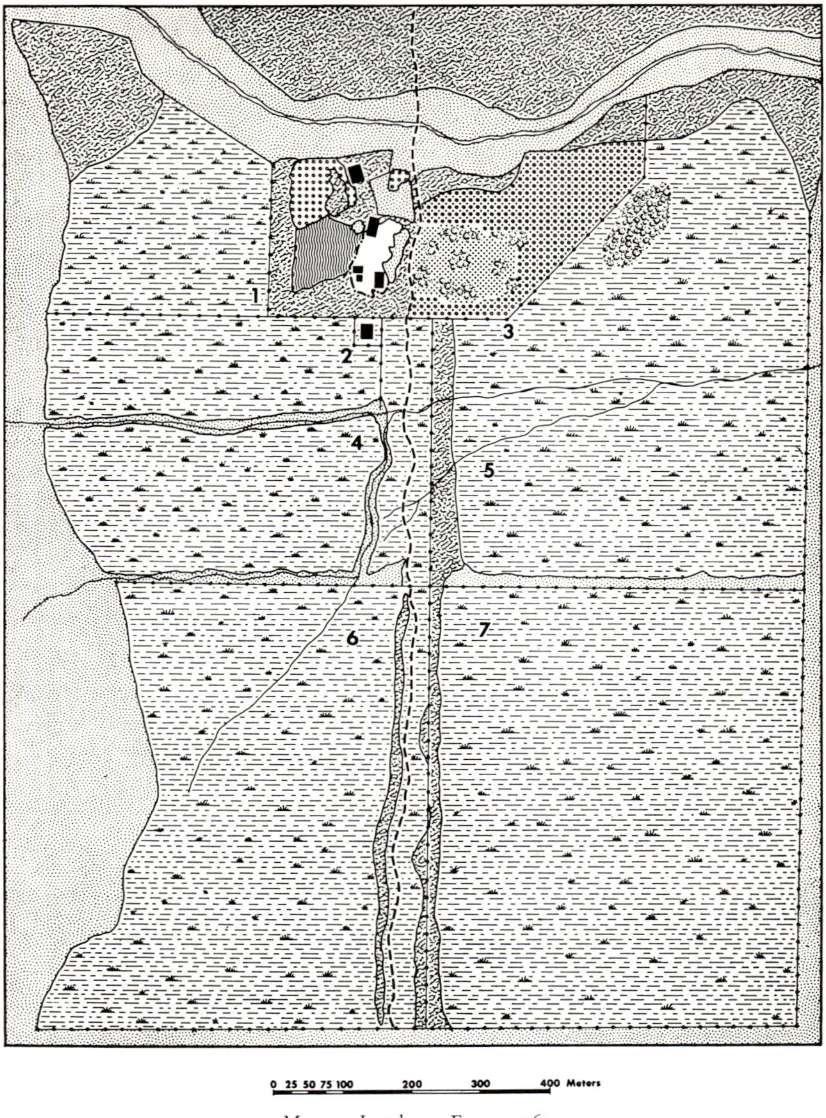

Map 11. Land use: Ewa—1969.

The final case described illustrates a partial system of shifting cultivation, in which the subsistence production plays only a minor role in newly cleared areas. The traditional system is now virtually impossible to distinguish except for the techniques of field clearance.

Evolution of the Finca. Ewa carved his *finca* out of unclaimed land bordering mission property in 1963, after he graduated from the El Tukuku mission school and married. Initially fields were cut alongside the Quebrada Kosíra using the standard shifting cultivation technique of the tribe. Working alone, he cleared the first field in the area now occupied by the house site, and pasture 1 (map 11). Maize, plantains and bananas, and pasture were sown together in the first year. During the first year of occupance a small area was partially cleared and coffee bushes were planted. The maize and plantains and bananas were harvested at the end of the first year, and the area left in pasture. Planting was extended outwards from the original nucleus during the second year, when maize, pasture grasses, manioc, and plantains and bananas were planted using traditional methods in two adjacent fields (pastures 2 and 3, map 11). In this manner the *finca* was extended by the end of 1967 to include pastures 2 through 5. In 1965, with the exception of the area alongside the house site, plantains and bananas were removed after only one harvest and the field remained in sown grasses. Pineapples were added in the western part of the house field, and fruit trees around the house site itself. Alongside the living site the plantains and bananas remained until the present, and have been used to provide shade for young coffee bushes. The newest pastures, 6 and 7, were cut and planted in 1968. Pasture 6 was sown in maize, manioc and pasture grasses, whereas 7 was sown in plantains and bananas, which, because of the hardness and dryness of the ground, did not yield. The plants were cut and burned and pasture was sown in 1969.

Finca Area

Pineapple	0.684 ha.
Plantains and bananas	0.656 ha.
Coffee with shade trees	1.514 ha.
Coffee, plantains and bananas	3.683 ha.
Pasture 1	7.204 ha.
Pasture 2	6.799 ha.
Pasture 3	12.667 ha.
Pasture 4	8.944 ha.
Pasture 5	13.274 ha.
Pasture 6	22.866 ha.
Pasture 7	33.347 ha.
Total	111.638 ha.

Livestock. Almost from the start of operations on this *finca,* the cultivator's intention was to raise livestock and cash crops, and not to bother much with subsistence production. In this respect he admits that he has been very impressed by what he has seen of the *haciendas* in the District of Perijá. He frequently expresses a desire to own one. During later years his subsistence crops were raised to feed only his labor force, whereas his family purchased their food in the Tukuku mission village from other cultivators.

During the second year of operation 25 pigs were purchased and kept penned on the holding. They were not permitted to forage and were fed cut pasture grasses, maize, and manioc. Later they were sold for 200 *bolívares* ($55 U.S.) each in Machiques, and cattle purchased with the proceeds. Currently the cultivator has 52 head of beef and dairy cattle of various ages and in various conditions. Pigs are no longer kept.

Formerly turkeys were kept in the *finca,* but losses to wild cats and foxes were so severe that they are now kept permanently in a fowl house in the mission village (see map 5). The domestic fowl kept comprise 30 chickens, two turkeys, and nine pigeons.

Labor. During the second year of operation it became apparent that additional labor would be needed to clear land sufficiently rapidly to set the holding on an economically sound basis. Accordingly two Irapa men and their families (5 persons in all) from the settlement of Sásape in the headwaters of the Río Tukuku were hired and settled in a house on the *finca.* For their subsistence they utilize some of the food crops grown on the land during its first year of cultivation. In addition, they are paid monthly with a combination of cash, clothes, and tools. Currently the cultivator is providing them with the materials to construct a brick-wall and zinc-roofed dwelling for themselves.

Productivity. Apart from the minor subsistence production of maize and manioc from first year fields, and plantains and bananas from the older fields, the production of the *finca* is entirely for sale. Surplus maize, manioc, and plantains and bananas are mostly sold to the mission, as are pineapples, fresh fruit, and coffee. Milk (24 liters daily in the wet season and 10 liters daily in the dry season) from the 7 producing animals is also sold to the mission. Cattle are sold for slaughter to the mission and in Machiques. Domestic fowl and eggs are consumed entirely by the cultivator and his family. The cultivator has a small store in the mission village, where he sells, in addition to a small surplus of agricultural products remaining after the bulk has been sold to the mission, household sundries, cigarettes, footwear, clothing, bread, candy, and soft drinks.

Life Style. Although biologically a Yukpa, this cultivator is to a large extent culturally a *wátia.* In reality his cultural loyalties are divided, for as a joint chief of the mission village appointed by the Capuchins, he is obliged to associate himself closely with Yukpa ways. Apparently he is able to move in both *wátia* and Yukpa worlds with equal confidence, and at the same time retain the esteem and trust of both parties.

Ewa, an Irapa, lives in the mission village with his wife, a Rionegrina, sister-in-law (who attends the mission school) and a young, unrelated Irapa male, who has recently come down from the Tukuku headwaters. This young man serves Ewa as a general factotum.

As befits his aspirations, his village house is quite atypical for a Yukpa,[12] with cooking and sleeping quarters separated by a well-maintained yard, with manicured lawns (see plate 17). The kitchen is equipped with a gas range, electric refrigerator and dining room set. The living quarters are tastefully furnished with bedroom and living room suites. Ewa rides to his *finca* on a late model Honda (disdaining the bother of keeping a truculent mule), where he has built and is furnishing a replica of his "town house."

Ewa's way of life is atypical of the average Yukpa, nor is it like that of the average Venezuelan and Colombian *campesino.* Indeed, he sees for himself a future as a *hacendado* and as an organizer and manager of a Yukpa cooperative in the lower Tukuku valley.

There is then a continuum of life styles and methods of living among the Yukpa. The continuum is rooted in the subsistence system which ranges from hunting and gathering activities with complementary shifting cultivation, to shifting cultivation with hunting and gathering as ancillary activities, to the most recent state, exemplified by Ewa, of cattle ranching and cash cropping, complemented by shifting cultivation.

Supply and Division of Labor

Labor is divided basically according to sex and age (table 11). It will be seen that the man's tasks are generally those which are either physically demanding or call for artistic skill. To the women are reserved the pedestrian, repetitive chores, and certain tasks, such as *mísk* sowing, which are prescribed by ritual. Old people who are physically incapable of undertaking the arduous tasks involved in the cultivation system or of partici-

[12] Many of those living in the mission village prefer to cook over an open fire in the backyard and to sleep on mats on the floor. The bed springs provided with the house make an excellent perch in the fowl house.

TABLE 11

SEXUAL DIVISION OF LABOR IN AUTOSUBSISTENCE ACTIVITIES

Task	Performers Men	Performers Women	Task	Performers Men	Performers Women
Field clearance	x		Collection of materials for bow and arrow making	x	
Míšk sowing		x			
Planting	x		Pottery making	x[4]	
Harvesting	x[1]	x	Basket weaving	x	
Weeding	x	x[2]	Utensil fabrication	x[5]	
Preparation of harvest for storage	x	x	Cotton preparation		x
Selection of seeds	x		Textile weaving		x
Hunting	x		Tumpline weaving		x
Fishing	x		Mat and fan making		x
(Gathering of fish)	x	x	Rope making		x
Gathering plants and insects, etc.	x	x	Food and beverage preparation		x
Transportation		x[3]	Household chores		x
Construction (house and other)	x		Firewood and water collection		x[6]
Medicine	x	x	Laundry		x
Bow and arrow fabrication	x		Government	x	x

[1] By tradition men help only occasionally with harvesting.

[2] According to Reichel–Dolmatoff (1945:73). I have heard only of widows cleaning fields in circumstances where the kinship obligations have broken down.

[3] Before the introduction of beasts of burden, women always carried the loads when travelling. This left the man free for hunting.

[4] Schön and Jam (1953*b*:73) stated that pottery making was done by both sexes, whereas Reichel–Dolmatoff (1945:36) stated that it was exclusively women's work. I photographed a woman making a pipe in the mission at El Tukuku. The photograph occasioned much surprise among the Yukpa men who said that women *never* make pottery.

[5] Schön and Jam (*op cit.*) noted this as women's work.

[6] Firewood collection was listed by Reichel–Dolmatoff (*op cit.*) as exclusively men's work. The men generally cut the wood, whereas the women drag it home.

pating in complementary autosubsistence activities, confine themselves to lighter activities.

Apart from the doctor and perhaps the potter, there appear to be no occupational specializations among the Yukpa. Every able-bodied adult is an expert at performing all of the tasks assigned by sex and age.

WORK GROUPS

The individual nuclear family comprises the basic work group for cultivation activities. Ancillary activities, especially hunting and fishing, necessitate the participation of extended family members, for both greater efficiency and ritual reasons. Communal activities, such as when the adult males combine for the final day of tree felling, are purely symbolic, and are aimed at emphasizing community solidarity. Traditionally, field burning involved exchange labor, but this too was in compliance with magico-religious dictates and does not challenge the independent work unit concept. In labor supply, as in all other facets of Yukpa life, traditional standards are now gradually disappearing, and instances of communal, wage, and exchange labor are becoming increasingly common.

Five basic arrangements of cultivation labor operate at present among the Yukpa. The first is that under the obligations of bride service and uxorilocal residence rules, a son-in-law is obliged to work either alone or with his father- and brother(s)-in-law on the father-in-law's land to support his wife's family. This labor arrangement is reciprocal in that the father-in-law helps the young man to work the latter's own land, should the need arise. Married men sometimes continue to help their fathers also, especially with the heavy work of felling. Thus, despite uxorilocal residence rules, a son's labor is not necessarily entirely lost to his father. Under the second arrangement, the head of the nuclear family works alone, or with his unmarried sons. This is encountered where uxorilocal residence and bride service have ended and the continuing obligations on the son-in-law have for some reason been flouted, or where the father-in-law has died. Third is the cooperative labor arrangement by which brothers share the work of cultivation and the resultant harvest. This, however, is not common. The fourth arrangement consists of reciprocal labor agreements. They operate mainly between cousins. The fifth arrangement, and one which is increasingly frequent among the Yukpa, is the newly adopted system of the recruitment of hired labor to help with the heavy work of field cutting and felling. Payment is made in numerous ways, the most common of which are: cash, clothes, and other non-traditional material goods, a share in the harvest, a specialized product, especially arrows, or a tract of land.

LOSS OF LABOR

When labor is lost owing to illness, injury, or death, the extended family insures that nobody will starve as a result. Other men often undertake the clearance and cultivation of an incapacitated man's land. Women, too, can do all of the sowing and planting, and call on male kin to help with the clearance phases of the cycle.

In the event of the death of a cultivator the institutions of levirate and sororate traditionally took care of the deceased's family members. Indicative of the breakdown in the Yukpa social structure are the cases noted by the author of two widows who were forced to undertake all of the cultivation labor, in the absence of any kinsfolk to aid them.

The Expenditure of Labor in Cultivation

Average man-hour costs per hectare are summarized in table 12. Only data for the major shifting cultivation activities are presented; related activities such as feasting, travelling, and house building are excluded. In some instances, journey to work times are in excess of three hours, but because my data on this topic are incomplete they have not been included here.

The data in table 12 are averages derived from timing work-hour activities during the various phases of the cultivation cycle, and noting the number of work days spent on each activity. The reader is cautioned that among the Yukpa the range of man-hour labor input is large, and the data were tabulated for cultivators who work consistently at a task. As may be inferred from the data presented below dealing with field sizes, a large number of Yukpa cultivators are occupied with both ancillary autosubsistence and non-traditional activities, such as general work in the missions, working on mission land, and working as hired hands on the farms of the *wátia*. Naturally the man-hour input of cultivation labor decreases proportionate to the increased input for the other activities. The seasonal distribution of labor, however, remains consistent and is independent of both field size and labor input.

The greatest amount of labor expended in any one year goes into clearing a new field. In the traditional Yukpa system, where the maize field is not weeded, and other fields are weeded at best only once or twice a year, the maintenance of fields after the first year involves only a minimal expenditure of labor, even in fields that were initially cleared in low regrowth. The average industrious work group, working fields cleared in tall forest fallow, expends a little in excess of 2000 man-hours per hectare annually in cultivation, of which about 870 are devoted to the maize crop, and about 1130

TABLE 12

ANNUAL AVERAGE CULTIVATION LABOR EXPENDITURE FOR FIELDS CUT IN HIGH FOREST

Activity	Man–Hours per Hectare (Traditional System)	Man–Hours per Hectare (Bean Field)
Slashing	135.9	135.9
Felling	333.6[1]	333.6[1]
Cleaning for re–use	111.2[2]	111.2[2]
Burning	4.9	4.9
Maize sowing	44.5	–
Interplanting (beans)	24.7	–
Bean sowing	–	44.5
Harvesting and storing, etc. (Maize)	358.3	–
Harvesting, etc. of Interplanted beans	160.6	–
Bean harvest, etc.	–	568.3
Other crops	741.3	–
Cleaning (2nd and 3rd year fields)	98.8[3]	–
Total	2013.8	1198.4

[1] Not applicable to fields cut in *rastrojo*.
[2] Add 50 percent for fields cut in *rastrojo*.
[3] Calculated on the basis of an individual cultivator working alone.

to the ritually less important but more laborious cultivation of non-grain crops, especially during the preparation and planting of the clones.

The addition of bean fields adds very considerably to the man-hour input of cultivation (table 12). For this reason very few Yukpa have adopted this newly introduced practice, because the majority still prefer to spend much of their time in hunting and other pursuits. However, the sale of surplus beans enables the cultivator to purchase a shotgun, which, to a large extent, compensates for the loss of hunting time by increasing his effectiveness.

Although tabulated cultivation data are useful as a summary, they have the disadvantage of obscuring more vital information than they reveal. For this reason it is worthwhile to illustrate labor expenditure with two brief case studies, in the same manner as the land use has already been examined. For the sake of continuity I will use two of the same families that were used to illustrate land use.

LABOR INPUT OF AMÚŠMO'S GARDEN FAMILY

Of the eight members comprising this garden family four contribute to the labor pool. Together they put in 3642 man-hours of cultivation labor in 1969. Such a gross figure obscures both the sexual division of labor and its annual distribution. The man-hour division by task and the work group members employed is illustrated in table 13.

The bulk of the labor, 1158 hours each (64 percent of the total), is undertaken by Amúšmo and his son-in-law alone. The total per worker female labor input in this case is 663 hours, of which maize sowing comprises only 34 hours. In the past the women alone undertook the *míšk* sowing, but now men also take part, and in this garden family, at least, they share the task equally with the two women. Harvesting, however, is exclusively the women's province, and constitutes 95 percent (629 hours each) of their total input.

Seasonal Distribution of Labor. The male labor input is concentrated in the early part of the cultivation cycle, between late November and April, which is a time of hard and often hectic work. Only a small percentage of the cultivation labor (manioc field clearing, and occasional replanting of the field) comes after this period.

During the slashing and felling phase of the cultivation cycle, Amúšmo and his son-in-law worked hard and continuously, from 6 a.m. until 4:30 or 5 p.m., except that the period of greatest heat, from 12 until 2:30 or 3 p.m., is spent resting.

In order to clear the area for the new 1969 maize field 93 days (of eight hours each) were required. Slashing and felling began at the end of November, 1968, and lasted until the first days of March, 1969. A further four days were required to clear an area in the previous year's field for planting manioc. This was done after the main cutting operation had been completed. With all the slashing and felling completed the cut material was left to dry out, and the whole of the garden family, in the company of others, left for ten days of fishing and relaxation in the lower Maracá valley.

TABLE 13

CULTIVATION LABOR INPUT: AMÚŠMO'S GARDEN FAMILY (1969)

Task	Area (hectares)	Man–hours per Worker	Total All Workers	Garden Family Members Employed	Percent of Total Input	When Done
Slashing and felling (1st year field)	3.162	744	1488	Amúšmo and son–in–law (2)	40.9	Dec.–March
Burning (reciprocal labor)	—	5	10	Amúšmo and son–in–law (2)	0.4	March
Maize sowing	2.979	34	136	Amúšmo, wife, daughter, son–in–law (4)	3.7	March–May
Interplanting in 1st year field	0.326	121	242	Amúšmo and son–in–law (2)	6.6	March–April
Cleaning manioc field	0.444	30	60	Amúšmo and son–in–law (2)	1.6	Feb.–March
Planting manioc	0.444	170	340	Amúšmo and son–in–law (2)	9.3	March–April
Weeding manioc	0.444	22	44	Amúšmo and son–in–law (2)	1.2	Variable
Cleaning to plant plantains and bananas	0.076	4	8	Amúšmo and son–in–law (2)	0.2	March–April
Planting plantains and bananas	0.076	28	56	Amúšmo and son–in–law (2)	1.5	April–May
Maize harvest	2.972	532	1064	Wife, Daughter (2)	29.2	Aug.–Nov.
Manioc, taro, and yautia harvest	—	97[1]	194	Wife, Daughter (2)	5.3	Jan.–Aug. (mainly)
Total	—	1158 (M) 663 (F)	3642	—	100.0	—

[1] Generally, in good years, manioc and yautia are used only to vary the diet when maize is available in abundance from August to January. Therefore the dietary role of root crops becomes more important, between January and August. Women are harvesting all year.

The fields were burned at the end of March, but because of the taboo on cutters burning their own fields, Amúšmo called on his nephew to burn for him. In the meantime Amúšmo and his son-in-law burned his nephew's fields. Burning occupied them for one day.

Sowing started at the end of the last week of March. Amúšmo, his wife, daughter, and son-in-law planted maize after the first few showers had dampened the earth and signalled the approach of the *tiprína ántasi.* The four sowers required a total of 136 hours to sow the area completely. To maintain varietal segregation, however, the actual labor input of maize sowing was spread out over a period of almost six weeks. For this reason sowing occupied the workers only for a couple of hours daily, except when larger blocks of a single variety were being sown. Coincident with maize sowing, the men were busy planting and sowing the other field crops. In all, during the period from the end of March until early May a total of 576 man-hours were consumed by cultivation labor. Of these 68 were contributed by the two women. The men were, therefore, fully occupied with cultivation labor for about six hours daily, the work lasting usually from about 6 a.m. until noon. The day's work done, Amúšmo usually went hunting during the late afternoon.

Under the traditional Yukpa system the annual cultivation labor of the male members of the garden family virtually ceased with the completion of sowing and planting. Occasional interplanting and filling-in of manioc, yautia, sweet potato, and plantains and bananas took up odd hours during the remainder of the year, but essentially their time was devoted to other autosubsistence activities. Under the influence of *wátia,* however, several days may now be spent weeding and cleaning second and third year fields.

Female and male cultivation labor input do not peak at the same time, the peaks being separated by a period of about two and one half months (late May until August) when no cultivation labor is undertaken by either sex. Female participation in cultivation is minimal during the time when the men are almost totally occupied with the cultivation activities. Apart from sharing in the sowing of maize their sole duty in cultivation, with the exception of the maize harvest, is to harvest root crops, a task which consumes an hour or less per day in the food-short season, if the fields are reasonably close to the homestead. When the new maize crop is ready, root crop harvesting is not undertaken. Female labor input peaks between August and November when the women put in long hours of tedious labor

TABLE 14

CULTIVATION LABOR INPUT: ČÁCO'S GARDEN FAMILY (1969)

Task	Area (hectares)	Man–Hours per Worker	Total All Workers	Garden Family Members Employed	Percent of Total Input	When Done
Slashing and felling (1st year field)	1.935	414	828	Čáčo and hired man (2)	35.6	Jan.–Feb.
Cleaning manioc field	0.778	86	86	Čáčo (1)	3.7	March
Cleaning plantain and banana field	0.071	12	12	Čáčo (1)	0.5	March
Burning (reciprocal labor)	—	5	5	Čáčo (1)	0.2	March
Maize sowing	1.247	19	57	Čáčo, wife and hired man (3)	2.5	March–May
Bean sowing	0.773	17	34	Čáčo, wife (2)	1.5	April
Interplanting in 1st year field	0.049	36	36	Čáčo (1)	1.5	April
Planting and sowing in manioc field	0.779	486	486	Čáčo (1)	20.9	Apr.–June
Planting in plantain and banana field	0.071	53	53	Čáčo (1)	2.3	Apr.–May
Maize harvest	1.247	193	386	Čáčo, wife (2)	16.6	Aug.–Nov.
Bean harvest	0.773	84	252	Čàčo, wife, cousin (3)	10.8	August
Manioc, taro, and yautia harvest	—	90	90	Wife (1)	3.9	Jan.–Aug. (mainly)
Total	Čáčo Wife Hired man Cousin	1405 403 433 84	2325		100	

harvesting and grinding maize in preparation for the seemingly endless rounds of harvest festivals and lunar month celebrations and during the final collection of the remainder of the crop.

LABOR INPUT OF ČÁCO'S GARDEN FAMILY

Unlike his uncle Amúšmo, Čáčo is no longer a traditional Yukpa cultivator, for as has already been shown, he is gradually incorporating cash cropping into his cultivation cycle.

Of the seven members comprising this garden family only two contribute to the cultivation labor pool. In order to provide sufficient labor to produce both subsistence and cash crops, Čáčo is therefore forced to hire additional help, and to call upon his cousin to aid him in times of particular need, such as during the bean harvest.[13] Together the work group put in 2325 man-hours of cultivation labor in 1969. In general terms (as can be seen from table 14), the labor input of this garden family, both quantity and seasonality, is similar to that of Amúšmo, with allowance for the incorporation of cash cropping and the reduction of female sowing labor which has resulted from acculturation.

PRODUCTIVITY

In western economic terms it is impossible to equate the productivity of the traditional Yukpa autosubsistence system entirely in relation to the cultivated surface area, because the subsistence system cannot logically be separated from the total life pattern. Disregarding the problematic question of the inclusion of the products of ancillary activities, the measurement of the productivity of the cultivated field alone is not a simple undertaking. Allowance has to be made for such important non-material factors as magico-religious and social dictates, which considerably reduce the amount of production available for consumption as food. As the traditional system becomes transformed under the influence of *wátia* ways, however, and other non-material elements of the culture are gradually lost, a more strictly economic analysis may be feasible.

The measurement of productivity is rendered more difficult by the complexity of the cultivation system in which many small, polycultural land units are utilized. The surface area of these small units was measured, but

[13] There was an acute danger of seed sprouting during August, 1969, owing to unseasonably heavy and continual thunder showers.

it was possible only to estimate their productivity, because reliable estimates for crops growing in various inter-cropped associations are not generally available. Allowance should also be made for loss and damage to both harvested crops and standing crops. In trying to account for damage to standing crops, it is difficult to incorporate Yukpa attitudes into western economic thought. To the Indians the loss of standing corn to predators represents not a "loss" but a "conversion" of the product: "instead of harvesting corn you harvest monkeys or squirrels." True, there is a point beyond which no more standing corn will be used as bait, but the precise location of this point depends on the inclination and palate of the individual cultivator.

Fieldwork revealed that there exists considerable variation in crop production figures per unit area among the Yukpa. Measures for the total yields of cultivated fields are not presented here, because sufficiently detailed data on the progressive decline of soil fertility, the variations in the micro-ecology of individual sites, differential yields of the varieties (of staple crops) and the variable yields resulting from the differential competition of various crop combinations were not obtained. Crop production figures from mixed fields are, therefore, only estimates.

On the basis of field measures among the Maracá and Irapa, the following productivity data for the stable autosubsistence crops were obtained:

Maize. Field measurements showed that one ha. yielded 1740 kilos of dried and shelled maize,[14] and since 877 man-hours were required to produce it, one man-hour of labor input was found to yield 1.98 kilos.

Manioc. Under the Yukpa system of cultivation the production of one ha.

[14] The productivity of the Yukpa system of maize cultivation compares favorably with that of other tropical cultivators. Watters (1965:16–17) estimated the maize production of typical Venezuelan *conuqueros* at 7968 kilos per hectare in the Andes and 919 kilos per hectare in the *tierra caliente*. Grant *et al.* (1963:7) estimated that a relatively low average yield of 1200 kilos per hectare was produced in 1958 in Anzoátegui, Cojedes, Falcón, Guárico, Portuguesa and Zulia, the leading maize producing states in Venezuela. Higgs, Kerkham and Raeburn (1950 cited by Clark and Haswell, 1966:81) estimated the "all-Africa" maize production to be 1790 kilos per ha., and Gourou (1958:16) gave figures of 1122 kilos per ha. for Mexico and 1871 kilos per ha. for Brazil, the Congo, and Indonesia. Tax (1953:50–52) estimates the production of the Panajachel Indians at Guatemala, working volcanic soils, to be approximately 2496 kilos per ha. Maize yields under systems of mechanized agriculture in the world's temperate region are considerably higher; Gourou (1958:16) cites yields of 3929 kilos per ha. for Argentina and 2993 kilos per ha. for the U.S.A.

of manioc, which yielded 4752 kilos,[15] required a labor input of 951 man-hours. That is, one man-hour of labor produces approximately 5 kilos (wet-weight) of manioc.

Plantains and Bananas. One hectare of second and third year fields required 951 man-hours of labor input (table 12) which yielded 6020 kilos of plantains and bananas (an estimated 81 stems,[16] the average weight of which was 30.1 kilos). That is, one man-hour of labor produces approximately 6.3 kilos.

Beans (not interplanted varieties). One ha. required 1087 man-hours of labor input (table 12) which yielded 3716 kilos of shelled and dried beans. Or, one man-hour of labor produces 3.4 kilos of beans.

Reliable estimates of the other crops were not obtained. As their importance in the economy of the garden family under examination was only minimal, the lack of data does not seriously alter the total production figures.

THE PRODUCTIVITY OF AMÚŠMO'S GARDEN FAMILY

	Available for Food:[17]	
Maize	2807 kilos	(312 kilos per capita),
Manioc	2807 kilos	(312 kilos per capita),
Plantains and bananas	4055 kilos	(451 kilos per capita),
Other minor crops	900 kilos (approx.)	(100 kilos per capita approx.)

Of the total maize production of 4897 kilos, 2090 kilos were diverted from direct food consumption in the following manner: 420 kilos were retained

[15] Statistical data for tropical root crop production are less readily available, hence it is more difficult to make meaningful comparisons with the manioc production of the Yukpa. Estrada (n.d.:1) estimated that Colombian production averages 6.54 metric tons per ha. Ramos (1970:83) notes that the Llanera variety yields 40 metric ton/ha. in the Cauca Valley of Colombia. This is an exceptionally high figure. He also states (ibid.) that this variety has yielded up to 70 metric ton/ha. on the experimental plots of the *Instituto Colombiano Agropecuario* at Palmira. In recent years the average yield in Brazil has been about 14 metric ton/ha., and that of the state of São Paulo 17.5 metric ton/ha. (Normanha, 1970:61). Manioc yield in India averages about 12.5 metric ton/ha. (Magoon, 1970:58). According to Jennings (1970:64) African yields are among the lowest recorded. Average production on that continent is only 6.7 metric ton/ha., compared with a world average of 8.4 metric ton/ha. (Normanha, 1970:61).

[16] For *criollos* Watters (1965:17) estimated the production of "bananas" on a "typical three hectare farm" in the *tierra caliente* to be 121 stems per acre.

[17] Amúšmo's garden family comprises nine "consumption units": 5 adults (1 unit each), 3 children (1 unit each), and 2 frequent guests who derive some of their subsistence from his production (0.5 units each).

for seed; 770 kilos were consumed in ritual; 900 kilos were lost (to corn borers and rodents).

The Yukpa system of maize cultivation, in which at least six grains are sown per hole, requires a large amount of seed corn. My observations and calculations showed that one kilo of maize seed was required to produce 25.66 kilos of harvested grain.

In addition, although this is now fast disappearing under the influence of the priests, a large percentage of the annual maize production is reserved for use in ritual. In order to produce sufficient food and liquor at the tripartite harvest festival for the whole of the Maracá subtribe, Amúšmo contributed 544 kilos of maize. Furthermore each garden family is obliged to contribute to the common pool for each lunar month celebration. This consumed a further 218 kilos of Amúšmo's production.

THE PRODUCTIVITY OF ČÁČO'S GARDEN FAMILY

The significant difference between the productivity of this garden family and that of Amúšmo arises from the fact that Čáčo is intent on entering the regional market economy, hence he has diverted much of his crop from ritual use to sales. The small quantity of maize available for consumption is compensated for by the abundance of beans. The total that is available for food to the garden family comprises:[18]

Maize	820 kilos	(164 kilos per capita),
Beans	2328 kilos	(466 kilos per capita),
Manioc	2885 kilos	(577 kilos per capita),
Plantains and bananas	1752 kilos	(350 kilos per capita),
Other crops	900 kilos	(180 kilos per capita).

Some of the total 1969 crop production was diverted in the following manner: Of the maize, 181 kilos were saved for seed, 900 kilos were sold, and 205 kilos were lost to rodents and borers. Of the beans, 225 kilos were used to repay a seed loan to the mission, 225 kilos were sold, and 90 kilos were saved for seed.

Given the confused and uncertain state of our knowledge about basal subsistence and occupational and nutritional requirements of tropical populations, it is difficult, especially for one untrained in the field, to be certain of the exact meaning of these data. To be sure, the implication, as the scanty

[18] Čáčo's garden family comprises 3 adults (3 "consumption units") and 4 young girls (2 units, because they derive a large percentage of their subsistence from mission school dinners).

medical data corroborate, is that the Yukpa are living at the lower subsistence margin, and that the products of shifting cultivation alone would not insure a balanced diet. The importance of the ancillary activities—hunting and insect gathering in the provision of animal proteins and fats, and vitamins provided by wild fruits and leaves—is immediately obvious.

Chapter 6

CONCLUSION

Throughout this study it has been stated and implied that there is neither a single Yukpa autosubsistence system nor, more specifically, is only one cultivation system practiced. It has been repeatedly emphasized, too, that the Yukpa life style and society is in transition, as the composite forces of *wátia* ways take their toll of the traditional culture traits.

Although all facets of Yukpa life are changing rapidly, in none is the change more fundamental than in the food quest. The traditional cultivation system is changing as new tools, plant and animal species, cultivation techniques, and land use patterns are adopted, and as new aspirations and totally new ways of making a living compete for the Indians' attention. Such changes have been mentioned in the relevant places in the text. I have, however, consistently and purposely ignored a significant and somewhat speculative question in the cultural history of the Yukpa; namely, the origins of the fundamental cultivation system utilized by the tribe. Piecing together the admittedly fragmentary and inconclusive evidence of legend, oral history, documentary history, and ecological observation, the development of the system may be outlined in a tentative manner.

The Hypothesized Evolution of the Traditional Cultivation System

Although the Yukpa manifest many of the traits characteristic of incipient tropical forest horticulturalists, they no longer completely fit into this category as a result of inferred changes, some of which may predate the Spanish Conquest, in their physical and cultural environment.

Before and during the early years of the Conquest, the Yukpa inhabited the lowlands which border the eastern flanks of their present habitat, within which time they appear to have been typical incipient tropical forest horticulturalists living in villages. At that stage in their history, shifting cultivation appears to have formed only one segment of their total subsistence system, together with the products of hunting, gathering and fishing, which constituted the major part of their food source.

The Conquest wrought major changes in the Yukpa autosubsistence system. The Indians were pushed out of the lowlands into the mountains, which at the higher altitudes offered only very meager resources for hunting, gathering and fishing. Furthermore, in the mountains, the productivity of the Yukpa's lowland system of shifting cultivation would have been reduced; and its technology, as evinced by contemporary activities, would have been extremely destructive and ill-adapted to the steeply sloping terrain of their new environment. Although the productivity of the total autosubsistence system must have been less, after a number of years when the effects of indiscriminate hunting made themselves felt upon the faunal population, its productivity may, judging by contemporary conditions, have declined much more rapidly than that of shifting cultivation. Similarly, the productivity of the other sectors of the mixed economy would have been greatly reduced by the shift from the lowlands to the highlands and also by the rapid destruction of the fragile ecosystems of the new habitat. Fish resources in the mountains are virtually nonexistent, and the forest products severely reduced compared to the lowlands. Furthermore, as with hunting, the pressure of population on the forest products may have severely reduced the supply, such that certain localities may have become virtually "collected out." Perhaps it was for this reason that a group of now feral plants were at one time planted, as a food conservation measure. These plants enter greatly into Yukpa discussions of the plant world, but are almost impossible to find now in the forest. It should be emphasized that I am not suggesting that a group of hunters, gatherers and fishers developed cultivation; but only that a group of incipient cultivators were forced into a new environment and sought new plants (either from contacts with other tribes or from the new environment) to help them to subsist and augment their tropical lowland crops, which did not produce well at higher elevations. I am postulating for the Yukpa a partial system of cultivation in which the role of shifting cultivation was expanded (and continues to expand) as the importance of the other subsistence activities declined. The structure of the society seems, however, to have made few concomitant changes.

Modern Cultivation Systems

Among the present day Yukpa the importance of shifting cultivation varies greatly. For some groups, hunting and gathering continues to play a major subsistence role, and the cultivated field remains secondary, whereas for others the roles are completely reversed. Some families depend completely on the products of the shifting field and undertake virtually no ancillary activities, whereas for others shifting cultivation has been replaced by cash cropping and cattle ranching.

The range of Yukpa partial systems of shifting cultivation parallels the typology established for shifting cultivation as a whole. Two basic partial systems are operated: The original system of the incipient tropical forest horticulturalists, and the adapted system with hunting, gathering and fishing relegated from coequal to ancillary subsistence activities.

More recently, other partial systems have begun to be employed by the Yukpa in which shifting cultivation is being relegated—and in one case completely dispensed with—as cash cropping and cattle ranching becomes more widespread among the subtribes.

Present Trends and the Future Yukpa Cultivation System

Assuming that the present demographic and cultural parameters remain constant, contemporary trends and recently adopted innovations within the Yukpa autosubsistence system portend future developments. As *wátia* culture traits reach into the interior of the region and affect the more isolated communities; and, too, as the isolated families descend to the newly emerging pivotal settlements of the subtribes, the modernization of the Yukpa, and the concomitant changes within their autosubsistence system, will proceed with increasing rapidity. Changes which are at present taking place in the system will be strengthened and will be adopted by the majority of the Indians.

Two basic changes are now taking place in the Yukpa autosubsistence system: First, the traditional ancillary activities are rapidly decreasing in importance, and are being replaced by ancillary activities characteristic of the Colombian and Venezuelan peasantry. Second, fundamental changes are taking place within the cultivation system, in terms of technology, crop assemblages, number of crops grown per season, and the cultivation schedule.

The Decline of Traditional Ancillary Activities

The Yukpa habitat, especially at higher altitudes, has, over the centuries of occupation, become "hunted out," and, to a lesser extent, "collected out." This has resulted not only from the Indians' hunting and collecting activities, but also from their careless and indiscriminate use of fire, which in many areas has drastically altered the ecology. Bird, animal, plant and insect species adapted to forest niches are now greatly diminished, and conversely those adapted to savanna, scrub, and fern associations, despite their increased populations, afford a poorer resource base.

Traditional ancillary activities, particularly entomophagy and anthropophagy, have declined drastically in recent years as a result of transculturation. Although eating invertebrates remains important dietetically in the more isolated communities (Ruddle, 1973) where it is more than a mere famine recourse, the trait is rapidly disappearing, as the modernized Indians, influenced by the *wátia,* regard it as an index of primitivity.

The traditional ancillary activities are also declining as alternative complemental occupations are adopted. Principal among these are cattle and other smaller livestock raising, cash cropping, and work as agricultural laborers in the missions and on the surrounding *haciendas* and *fincas.* In short, the Yukpa are changing from an ancient to a modern system of partial shifting cultivation.

Future Prospects

The modernization of the autosubsistence system of the Yukpa seems bound to continue at an increasing rate. Under the protection of the missionaries at the El Tukuku mission, the Irapa are planning their moderization via a small cooperative designed to handle livestock, and both cash and subsistence cropping. The subtribes beyond the effective control of the priests are haphazardly adopting the techniques of their *campesino* neighbors.

Judging by the attitude of many of the younger Yukpa at present attending the mission school at El Tukuku, the number of future cultivators will be greatly reduced as many seek employment and a life beyond the reserve. These who intend to remain in the area increasingly aspire to raise cattle, coffee and other cash crops, and to enter the regional market economy. This presages major changes in the cultivation system, which may within the next few decades include only remnental shifting characteristics. There is a very marked trend toward sedentarized agriculture among the most modernized groups. The other subtribes, as has been noted, given the continuance of present trends, may soon cease to exist as separate entities.

Appendix

MARACA ORTHOGRAPHY

Preliminary Observations

The Maracá subtribe speaks a dialect of Western Carib, a language composed of several dialects which are mutually intelligible in the Sierra de Perijá of Venezuela and Colombia. These dialects include, on Venezuelan territory, Macoita-Rionegrino, Parirí, Wasama, Shaparrú, Irapa, and Viakši. On the Colombian side they comprise Iroka and Yuko. Japreria is a closely related but mutually intelligible dialect (language?) spoken in the northern part of the Sierra in both Venezuela and Colombia.

Although there are many publications of various levels of sophistication which date as far back as the early colonial period and which contain word lists of varying lengths, there is only one recently published linguistic treatise available for Western Carib (Hanes, 1952). Hanes' work deals with the Iroka, who are geographically near but not, apparently, linguistically close to the Maracá. He lists a set of nasalized vowels and a retroflexed /č/ for Iroka which Maracá does not seem to have. Hanes does not, however, show /r/ which Maracá does have. Maracá, as gathered by Dr. Ruddle in the field, seems to be closer to the Yukpa dialects on the Venezuelan side than Hanes' Iroka on the Colombian side.

Orthographic Conventions

The compilers of this glossary have not had experience with the Maracá dialect. The following orthographic conventions are based on Dr. Ruddle's phonetic notes. Our observations are based on fieldwork conducted in 1972 with the Yukpa dialects on the Venezuelan side of the Sierra (Durbin and Seijas, 1972). Consequently, our observations are at best educated guesses, although we believe them to be essentially accurate. This belief is based largely on the fact that Dr. Ruddle has a keen ear for phonetic details and has accurately recorded these details in his prolonged work in the area.

The Maracá dialect has six vowels (table 1).

APPENDIX TABLE 1

Maracá Vowels

Tongue Height	POSITION OF THE TONGUE			
	Front	*Central*	*Back*	
			Unrounded	*Rounded*
High	i		?	u
Mid	e			o
Low		a		

In addition to the above, Dr. Ruddle has also recorded a few cases of lengthened vowels (aa, ii, uu). The other vowels appear not to be lengthened. This contrasts with Hanes' Iroka, where all vowels are lengthened, and with the dialects on the Venezuelan side, where no vowels are contrastively lengthened. Dr. Ruddle has also recorded several examples of one diphthong (ia). It is quite possible that both the lengthened vowels and the diphthong are cases of VOWEL + y + VOWEL. However, lacking further evidence we have left Dr. Ruddle's transcription as he recorded it.

The high back unrounded vowel (ɨ) is the one commonly written as ë or ï and appears to have two free variants in Maracá, [ï] and [Ï]; that is, higher and lower variants, respectively, though the former is very infrequent in Dr. Ruddle's data. Following our own knowledge of these dialects, the other vowels would have the following values: (i) as in Spanish nido, *nest;* (e) generally as in English *bet,* though it may vary with Spanish pecas, *freckles;* (u) as in Spanish puma; (o) as in Spanish tomo, *I take;* (a) as in Spanish padre, though it may vary with English *but.* Words may begin and end with vowels; and, as mentioned above, vowel clusters are very infrequent.

The consonants are given in Table 2.

APPENDIX TABLE 2

MARACÁ CONSONANTS

POSITION OF THE TONGUE

Manner of Articulation	*Labial*	*Dental-Alveolars*	*Palatal*	*Velar-Pharangeal*
Unaspirated Voiceless Stops and Affricates	p	t	č	k
Fricatives	v	s	š	h
Voiced Nasals	m	n	ñ	
Voiced Lateral (Flapped)		r		
Semi-Vowels	w		y	

The consonant (t) has two principal variants, [t] and [d]. The latter occurs when followed by (r). (t) may be unreleased [t̚] in final position and rearticulated [t͡l t] before (r). (c) also may be rearticulated [t͡l š] intervocalically and in final position. (v) has two voiced variants, [b] in initial position before (o) and in medial position; [v] occurs elsewhere mainly before (i) and (e). (ñ) appears to have two free variants [ñ] and [ny]. All other consonants seem to have only one variant similar to the description given in table 2.

Maracá has a large number of consonant clusters both in syllables and in morphemes. This is not so much the case in the dialects on the Venezuelan side of the Sierra, but it is characteristic of Iroka, which may have up to six consonants in sequence (Hanes, 1952:148).

There are two semivowels: (w) and (y). Maracá stands apart from the other dialects in this respect since (w) and (y) are not usually distinct in the other dialects. In Maracá, however, we find viš, *tree, stick, wood* vs. wičo, *sun* and voripa, *woman* vs. woši, *kind of grass.* In spite of the fact that (v) and (w) may be variants in Maracá, we have left them as separate contrasts based on the above information. (w) and (y) both occur in initial, medial (following vowels) and final positions. (y) is geminated in intervocalic positions.

There appear to be at least two degrees of significant stress in Maracá (strong and weak, with only the former being marked [′]). There are undoubtedly distinctive intonation contours as well but we have left them unmarked pending more information. Hanes (1952:147) notes that in Iroka "tone or intonation may be phonemic, but at present there is insufficient data regarding this." Dr. Ruddle marked three degrees of stress in his data and we suspect that one of these refers to intonation. Primary stress (′) often falls on the penultimate syllable of a word but not predictably so.

The following transcription is "phonemic" as nearly as we can interpret Dr. Ruddle's excellent phonetic transcription. For accurate pronunciation the above rules would be employed.

The names of cultivated, feral and wild plant species are not included in the glossary. The translations are free rather than literal. They should be understood as paraphrases of the observations Dr. Ruddle made in the field with the Maracá subtribe and those groups of the Irapa subtribe (such as the Pišíkakaw and Sásape) who inhabit the headwaters of the Río Tukuku.

When it was possible, we supplied literal translations and cross-referencing to other entries similar in form and meaning.

The alphabetical order is: a, č, e, h, ɨ, i, k, m, n, ñ, o, p, r, s, š, t, u, v, w, y. Note that no word in this glossary begins with h, ñ, r, or u.

Marshall Durbin
St. Louis, Missouri
Haydeé Seijas
Caracas, Venezuela

Glossary

A

akiríti. Grapnel used in conjunction with the machete for clearing fields and weeding crops.

ámor sapóma (literally, you who smoke). One who burns dry slash in a new field.

ámor šatámas (literally, you who are coming). To drive game with fire.

ámuk. Slash (i.e., undergrowth and *rastrojo* small enough to be cut with a machete).

apóte. Sleeping mats. Also mats used for drying harvested crops.

apóte intášpoš (cf. *apóte,* literally, mat for drying). To dry corn in the sun.

arápa. Loft in dwelling house.

árkna sayáta menúri (cf. *menúri*). The supernatural power exerted on maize by the collecting basket during the *túkapo osimáytpo* celebration.

atánto. The spirit of the sun.

atáyana. Tributary of stream.

awɨpranónošk (cf. *nóno, -šk*). Fallow field (i.e., area of *rastrojo*).

Č

čaánkano. Waterfall.

čánakarašk mokóšk (cf. *-šk, yémokor*). Suckers of *Canna edulis* used for planting.

E

etáka. To harvest. E.g., *etáka póyši,* to harvest manioc (cf. *póyši*).

Ɨ

ɨpɨymano. House site.

I

ikanmɨráktay (cf. *ikantirátay* below). A malevolent spirit.

ikáno mánponot. A malevolent spirit.

ikantɨrátay. A benevolent spirit.

imánta. The rain spirit.

imántape (cf. *imánta*). Hail that comes with the *tiprína ántasi*.

ípčane (cf. *ipčápo*). Withered maize leaves.

ipčápo (cf. *ípčane*). Corn plants dessicated because of insufficient moisture.

ipíḳane. Marshy ground.

K

ḳamá ántase. Trees felled to fall downslope.

ḳanówa. Trough in which *chicha* is prepared.

ḳaráčašḳ (cf. *-šḳ*). Crib for storing shelled maize.

ḳarína munɨ́ra (from Spanish *gallina,* cf. *mɨ́na*). Chicken coop.

ḳosípo osimáytpo (cf. *osimáytpo*). First part of harvest festival.

ḳosíš. A food known to the Venezuelan *criollos* as *hayaquitas.* They consist of ground maize wrapped and boiled in the husk to conserve the original shape of the cob.

ḳoštríča. Tunic worn by men.

ḳúna. Water.

ḳúnanane (cf. *ḳúna*). Drought.

ḳúnananɨ pánonim (cf. *ḳúna, ḳunáyunay tɨ́tnim*). Plants felled by soil wash-out around roots.

ḳúna tátpo (cf. *ḳúna*). Landslides and slumps.

ḳunáyunay tɨ́tnim (*ḳuna* = water, cf. *ḳúnananɨ pánonim*). Soil erosion.

ḳurátrɨt (cf. *ḳúratšḳ*). Palisade around house.

ḳúratšḳ (cf. *ḳurátrɨt, -šḳ*). Cleared ground between house and palisade.

ḳurɨ́ḳa. Generic term applied to larvae, such as that of *Mocis repanda F.,* which damage maize.

ḳuríḳa yɨpóḳšɨ (cf. *ḳurɨ́ḳa*). Corn damaged by worms.

ḳurúḳmatoro. Chicha server made from ligneous spathe of the *Sheelea* sp. palm; palm of *Sheelea* genus.

M

máḳ šúrus (*šúru* = skin, covering). Corn husk.

manú. Deadfall for trapping rats and other small rodents.

mapíča. Beeswax mixed with the ashes of various plants; a stunning arrow with head of beeswax.

marípta. Low forest fallow (*rastrojo*).

mašít (from Spanish *machete*). Machete.

mášuḳ ápa wátpo (*wátpo* = big, chief, head, leading). Male birth celebration.

matríčay. First fruits celebration; immature ears of corn; play arrows.

mayíša (cf. *máysiw*). Varieties of *Zea mays* introduced by the *wátia.*

máysiw (cf. *mayíša*). Corn cob.

menúri. Woven basket.

menúri súwaḳ (cf. *menúri*). Men's dance during the *túḳapo osimáytpo* celebration.

mɨ́na. Permanent dwelling house.

mɨ́šeḳ (cf. *mɨ́na*). Small temporary house.

mimíšk (cf. *-šk*). Vine, liane.

mimitakrɨnkas (*cf. mimíšk*). To cut lianes when slashing.

mišayñatpo. Basket in which sprouted *míšk* seed is carried to field for sowing.

míšk. Traditionally cultivated varieties of *Zea mays*.

míšk inuš (cf. *míšk*). Seed corn.

míšk tépese (cf. *míšk*). Maize flour.

míšk timási (cf. *míšk*). To sprout seed *míšk*.

N

nánta. Premature snapping of trees during felling.

naškátšk (cf. *-šk*). Holes for sowing annual crops.

natápa. Wet season.

nɨkóta. Slashed vegetation which, when dry, forms a flammable "bed" beneath the felled ligneous material.

nóno. Land, soil.

nóno kuráčašk (cf. *nóno, -šk, kuráča* = black). "Black Earth."

nónoma (cf. *nóno*). Bare earth swept clean in house compound.

nóno pahtúme (cf. *nóno, pahtúme* = good, beautiful). Cropland.

nóno sɨwɨswɨkano (cf. *nóno*). "White Earth." Any deeply leached type of soil which is useless for cultivation.

nóno tínkas (cf. *nóno*). Hole dug in soil for temporary storage of harvested root crops.

nonúkutk (cf. *nóno.*). Hole dug for planting banana and plantain suckers, and trees.

O

okátu. The malevolent spirits of dead Yukpa.

ompómšk (cf. *-šk*). The Yukpa assemblage of cultivated plants.

ontrókut. Hole dug for planting manioc.

orótkat. Group of houses; extended family dwelling compound.

osimáytpo. Harvest festival celebration consisting of *kosípo osimáytpo, túkapo osimáytpo,* and the *soñápo osimáykitpo*.

otópay iránowe. The overlapping crowns of felled trees.

owáya. A first year field (or maize field).

owáya tɨni (cf. *owáya*). An intended first year field which is accidentally burned, but which is, nevertheless, plantable.

oyanátkut (cf. *oyánonoke*). Confluence of streams; intercropping of plants (semantic extension of preceding).

oyánonoke (cf. *oyanátkut*). Entanglement of tree crowns which occurs during felling.

oyɨta mosmɨna (cf. *mɨna*). Store house and temporarily occupied field house.

oyúr pumána. Taboo involved in *túkapo osimáytpo*.

oyúwirpamasi. Taboo which forbids field cutter to burn his own field.

P

pánatarik̨. Stumps of felled trees.

paráyape. Very moist clayey soils.

pášk̨ (cf. *-šk̨*). Clay cooking pots.

pɨhpa. Fan.

pirápirača (*píra* = hard). Very hard, fertile, and exceptionally black class of soil.

piríko san sintrókak̨ (*piríko* = *Panicum miliaceum*). Strengthening of maize stalks by sympathetic magical powers of the *piríko* stem.

pošik̨átpo. Stick formerly used for harvesting manioc.

póya. Manioc field.

póyši. Manioc (*Manihot dulcis*).

póyšiwor (cf. *póyši*). A manioc cutting suitable for planting.

pšamtɨ́šk̨ (cf. *-šk̨*). Birdblind.

pšášk̨ (cf. *-šk̨*). Bottle gourd.

S

sak̨opátpo. A rite performed before starting the felling of trees.

sak̨rɨ́nk̨a. Preliminary cutting of undergrowth before felling.

samok̨ápu. Slashing low growth with machete during initial field clearance.

sápara. Bow string or any rope or string made from fibers of *Agave viridis* ssp.

sapóčpo. Special type of *k̨osíš* used at the *matríčay* celebrations; name for thread used to tie the *k̨osíš* which is worn around wrist by men who will prepare the next season's field. It is worn until field clearance starts anew.

saríkpa (cf. *saríkra*). A cultivated field, cultivation.

saríkra (cf. *saríkpa*). Unclaimed land.

sásare (sand). A sandy, alluvial class of soil.

sɨ́čk̨apɨ́ (cf. *sɨ́čk̨as*). An area of cut and drying vegetation destined for use as new field site (refers to condition of site and not to drying vegetation).

sɨ́čk̨as (cf. *sɨ́čk̨apɨ́*). To "clean" a weed infested field with a machete.

sɨriášk̨ ik̨átpo (cf. *-šk̨*). Stick formerly used for harvesting yautia.

sk̨opéta (from Spanish *escopeta*). Shotgun.

sóña. Fermented *chicha.*

soñápo osimáyk̨itpo (cf. *sóña, osimáytpo*). Third part of harvest festival.

sorúmspa. Large gourd.

súrik̨asa. Primary firing of the fields.

Š

šak̨rátɨtɨ. Leaf-cutting ant (*Atta* sp.).

šámpasa (cf. *šámpi*). Grains from thicker part of maize ears which are used for sowing.

šámpi (cf. *šámpasa*). Large basket with a capacity of about 45.5 kilos of grain.

šára. Fruit of *Poulsenia armata*.

šára tápoymotos (cf. *tapóymotos*). Ladder constructed for harvesting *šára*.

šára tɨrkás. Cold drink made from the fruit of the *šára*.

šɨ́trah. Digging stick.

šikárto. Hunting platform.

–šk. A suffix generally attached to nouns indicating diminutiveness, smallness, or endearment. The affix is very frequently used in Maracá speech.

T

tamúyas. "Packets" of maize flour used to make *sóña*.

tamúyas yipónš. Racks for drying *tamúyas*.

táñas. To sow maize, plantains, bananas, and papaya.

tápara. Rack constructed over fire for smoking meat.

tapótritos. Trellises for training beans.

tapóymotos. Platform constructed for cutting trees above buttresses.

tášpos. To sow tobacco and pigeon peas.

tátrisi. To cut tree above buttress.

táya. A seed.

tayíto. A malevolent spirit.

táyñapra. Part of *tɨ́šrane* which is permitted to return to fallow.

téva. Palm of *Guilielma* sp. Name given to wood of this palm which is used for making bows and some arrowheads.

tɨnarímas. Filling in planted field; to replant.

tɨ́nis opóčpo (cf. *sapóčpo*). Post field-burning ritual performed by women.

tɨ́ntpip. Slopes which are almost vertical.

tɨ́pomas (cf. *tɨ́pomo*). To plant taro, squash, calabash, tomato, red pepper, sugar cane, and yautia.

tɨ́pomo (cf. *tɨ́pomas*). To sow.

tɨpóta pnóya. Wind-damaged crop.

tɨ́ra. Part of *tɨ́šrane* which is planted.

tɨ́šrane. Field in which the felled vegetation fails to burn.

tɨtátkas. To shell beans.

tiíčas. To plant manioc.

tímsaš nimápinini. Maize damaged by activities of birds and mammals.

tínkas. To shell maize.

tipértartos. Crib for storing unshelled maize.

tiprína ántasi. "*Chichara* singing"; the rains which fall in April and precede the rainy season proper.

tíšiisip. Property line.

títrano. Land ruined by escaped burns.

tokántoksi. Stone knife formerly used to harvest sugar cane.

tokíru. Wet pit for preplanting taro.

tokórši. To rearrange logs for re-burning.

tókpa. Piles of dried leaves made before field is burned.

tókru. Land rendered uncultivable because of rockiness.

tomáčkas. Opening cut in *yɨtáprɨr* in order to serve *chicha*.

tómark surîkas (*tómark* = two). Secondary and tertiary re-burning of cleared field.

tómayra. Priest-shaman in charge of Yukpa cult life.

tóp. Rock (also basal stone in quern); stone; sharpening stone.

tópanas. To pull up weeds with hands.

totíkas. To burn weeds.

towápi tɨ́ni. A *sɨčkápɨ* destroyed by an escaped burn.

towára wɨ́tɨ oyáya (*towára* = all, complete; *wɨ́tɨ* = fire). An escaped burn.

túčkas. To weed a field.

túka. Unfermented *chicha*.

túkapo osmáytpo (cf. *osmáytpo*). Second part of harvest festival.

túka yapúsipi. Ceremony at start of tree felling.

túka yipákpa sitáka. Ceremony at start of tree felling.

túksi. Mounds of logs cut for re-burning; lid of *chicha* trough.

tuwánco (*tuano*). "He who knows"; shaman whose principal function is that of medical doctor.

túwate šínin. Weed infested field.

tuwíča. Tall secondary forest (high forest fallow).

tuwítarapas. Bird-bait stands.

tuwísarka. Crop row (preceded by name of crop).

V

víaritpo. Stone axe.

viartíkas (cf. *víaritpo*). Technique for multiple felling of trees.

víkan tótoma. Pointed stick formerly used for harvesting plantains and bananas.

vípopa. A sandy soil low in fertility.

viraskopákos (*virásko* = smoke). Smoking or those maize and millet seeds to be used for sowing in order to reduce damage by borers.

viruúsa. Ash.

viš. Live tree, stick or piece of wood.

višípra. Post-dibbling taboos observed by those women who have sown *míšk*.

vorípaš tomásay (*vorípa* = woman). Post-dibbling taboos observed by women who have sown *míšk*.

W

wakára. Bow.

wakára kákšk (cf. *wakára*, *-šk*). Small bow.

wanápsa. Tribe that inhabited the mountains before the arrival of the Yukpa.

wáneķ. Tumpline.

wánopo. Beeswax.

wápia. Fern associations and land dominated by them.

wáša. A steel axe.

wášaķan tomáķare (cf. *wáša*). To fell trees with an axe.

wátia. Any person who is not biologically an Amerind.

wayíķu. Clay used to make ceramics.

wɨrápa. Unripe maize.

wíčo. The sun.

wíčo siķánpese (cf. *wíčo*). Excessive drying of young plants owing to high evapo-transpiration rates.

wiķáno. Flattened digging stick formerly used for planting root crops.

wíta. All slash and felled material in field when cut.

witimímo. Protective charm placed over *sóña* trough.

wíto. Gourd and calabash.

wiyɨyaķátpo. Celebration at end of field cutting.

wopɨķš. Millet seed used as bird bait.

wóši. Savanna grasses and tracts of land dominated by grass associations.

Y

In the section which follows some of the entries beginning with (y) followed by a vowel may have the meaning "its . . . ." (E.g., *yíšit* literally means *its roots, its* referring, of course, to the tree.)

yáķa. Coleoptera of the Curculionidae family (corn borers).

yáķan nimápinini (cf. *yáķa, tímsaš nimápinini*). Corn damaged by borers.

yanúš. Cormel of taro and yautia.

yápoymor. Sapling.

yayórwort. Circular mound heaped up for manioc cultivation.

yémoķor. Suckers of plantains and bananas used for planting.

yéro. Dry season; a prolonged period of dry weather.

yɨtáprɨr. Airtight cover of banana and plantain leaves made for *sóña* trough.

yíķorat. Buttress of tree.

yimáķpit túnķus. To girdle trees.

yimáror. Corm of taro and yautia.

yimópe. Father or any authoritarian figure; one of the posts erected to guard maize in the storage cribs located in the field house.

yinípríšķ (cf. *-šķ*). Egg.

yipáķput. To pollard; tree trunk.

yipáķusumiķu. Firebreak around field.

yipɨ́nin antási (cf. *tiprína ántasi*). Taboo against premature drinking of *tuķa*.

yišámomo. Gullying

yíšit. Root.

yóčɨrɨt. Basal stone for grinding stones.

yooč nátɨnk̜os. First cut in tree felling.

yoonátrɨnk̜os. Second cut in tree felling.

yopáyet. Firewood.

yopótšk̜ (cf. *-šk̜*). Handstone for grinding stone.

yórsatɨ. Guardians of animals.

yówa. Ridge.

yuk̜áva. A malevolent spirit.

yunápčašk̜ (cf. *-šk̜*). Sweet potato and yam tubers suitable for planting.

yuwášač. Tip of maize ear.

yuwičɨr. To pull up weeds when sowing.

yuwówašač. Maize grains used in *tɨnis opóčpo* ritual.

BIBLIOGRAPHY

MANUSCRIPT SOURCES

Archivo Nacional de Colombia, Bogotá,
Conventos, t. 15 f. 548.
Conventos, t. 32 f. 253.
Conventos, t. 32 f. 255.
Curas y Obispos, t. 20 f. 814.
Eclesiásticos, t. 12 f. 414.
Historia Eclesiástica, t. 15 f. 260.
Milicias y Marina, t. 121 f. 121.

Archivo del Vicariato Apostólico de Valledupar, General Papers.

Cintruénigo, Fray Mauro de,
Archivo General de las Indias, Seville, *Santafé*, 547.

PUBLISHED SOURCES

ALCÁCER, ANTONIO DE
1959. *Las misiones Capuchinas en el nuevo reino de Granada hoy Colombia (1648–1820)*. Puente del Común (Cundinamarca): Ediciones Seminario Seráfico Misional Capuchino.
1962. *El indio Motilón y su historia*. Bogotá: Ediciones Paz y Bien.
1965. *En la Sierra de Motilones*. Bogotá: Ediciones Paz y Bien.

ALLAN, W.
1967. *The African Husbandman*. New York: Barnes and Noble.

ARMELLADA, CESÁREO DE
1946. La vida misional en "El Tukuku." *Venezuela Misionera*, 95: 368–372.

BAILEY, L. H.
1969. *Manual of Cultivated Plants*. Revised edition. New York: Macmillan.

BARRAU, JACQUES
1958. *Subsistence Agriculture in Melanesia*. Bernice P. Bishop Museum Bulletin (Honolulu) 219.
1961. *Subsistence Agriculture in Polynesia and Micronesia*. Bernice P. Bishop Museum Bulletin (Honolulu) 223.

BARTLETT, H.
1956. Fire, Primitive Agriculture and Grazing in the Tropics, In W. Thomas, ed., *Man's Role in Changing the Face of the Earth*. Chicago: University of Chicago Press.

BOLINDER, GUSTAF
1958. *We Dared the Andes*. New York: Abelard–Schuman.

BOOY, THEODORE DE
1918*a*. The Western Maracaibo Lowland, Venezuela. *Geographical Review*, Vol. VI, No. 6: 481–500.
1918*b*. An Exploration of the Sierra de Perijá, Venezuela. *Geographical Review*, Vol. VI, no. 6: 385–410.
1918*c*. The People of the Mist. *The Museum Journal*, Vol. IX, no. 3 and 4: 183–224.

Bowden, J.

1956. Maize Stem Borer Control Extension Dusting Trials First and Second Seasons, 1954. *Ghana Farmer,* 1: 23–26, 59–61.

Bristol, Melvin L.

1969. Tree *Datura* Drugs of the Colombian Sibundoy. *Botanical Museum Leaflets,* Vol. XXII, no. 5. Cambridge: Harvard University.

Brown, W. H.

1919. *Vegetation of Philippine Mountains.* Manila.

Carneiro, Robert

1960. Slash-and-Burn Agriculture: A Closer Look at Its Implications for Settlement Patterns. In Anthony F. C. Wallace, ed., *Men and Culture.* Selected Papers of the 5th International Congress of Anthropological and Ethnological Sciences (Philadelphia, Sept. 1–9, 1956), pp. 229–234. Philadelphia: University of Pennsylvania Press.

1961. Slash-and-Burn Cultivation Among the Kuikuru and Its Implications for Cultural Development in the Amazon Basin. In Johannes Wilbert, ed., *The Evolution of Horticultural Systems in Native South America: Causes and Consequences.* Antropológica Supplement, Publ. no. 2: 47–68. Caracas: Sociedad de Ciencias Naturales La Salle [S.C.N.L.S.].

Champetier, de Ribes, G., et al.

1963. Geologia y mineralizaciones cupriferas de la Serranía de Perijá, entre Becerril y Villanueva (Departamento del Magdalena, Intendencia de la Guajira). *Boletín Geológico* (Bogotá), Vol. XI, no. 1–3: 133–188.

Chaves Vargas, Luis F.

1962. *Sistemas agrarios en la producción de cereales, granos leguminosos, raices, tubérculos y musáceas en la región de los Andes.* Caracas: Ministerio de Agricultura y Cria, Dirección de Recursos Naturales Renovables.

1963. *Geografía Agraria de Venezuela.* Caracas: Universidad Central de Venezuela, Ediciones de la Biblioteca.

Clark, Alexander L.

1938. *The Feast of the Yowari (Notes on the "Out-Burial," as Practised Still Among the Venezuelan Indians—Motilone-Aruak—of Perijá, Zulia).* Maracaibo.

Clark, C., and Haswell, M. R.

1966. *The Economics of Subsistence Agriculture.* Second edition. London: Macmillan.

Conklin, H. C.

1954. An Ethnoecological Approach to Shifting Agriculture. *Transactions of the New York Academy of Sciences.* Ser. II. 17, 2: 132–142.

1957. *Hanunóo Agriculture: A Report on an Integral System of Shifting Cultivation in the Philippines.* Forestry Development Paper, no. 12. Rome: United Nations Food and Agricultural Organization (F.A.O.).

1963. *The Study of Shifting Cultivation.* Studies and Monographs, VI. Washington, D.C.: Pan-American Union.

CRIST, RAYMOND E.
1969 Human Geography and Neighboring Disciplines. In *Interdisciplinary Relationships in the Social Sciences*. Chicago: Aldine Publishing Company.

CUTTING, C. V., WOOD, R. A., BROWN, P., and AMBROSE, H. B.
1957. *Assessment of Fertility Status and the Maintenance of Productivity of Soils in Nyasaland*. Third Inter-African Soils Conference, Paper no. 69. Dalaba.

DIAZ-UNGRÍA, A. DE
1971. Microevolución en las poblaciones indígenas Yupa: Datos preliminares. Conferencia presentada en la Convención de la Asociación Venezolana para el Avance de la Ciencia. Capítulo Zuliano, Maracaibo. Mimeographed.

DUPOUY, WALTER
1958. Noticias de los indios Japreria. *Antropológica*, 4: 1–16.

DURBIN, MARSHALL, and SEIJAS, HAYDEÉ
1972. The Phonological Structure of the Western Carib Languages of the Sierra de Perijá, Venezuela. In *Proceedings of the International Congress of Americanists*. Rome: in press.

ERNST, ADOLF
1887. Die Sprache der Motilonen. *Zeitschrift für Ethnologie*. Berlin, 19: 376–378.

ESPINAL, LUIS SIGIFREDO, and MONTENEGRO, ELMO
1963. *Formaciones vegetales de Colombia*. Bogotá: Instituto Geográfico "Agustin Codazzi."

ESTRADA, NELSON RAMOS
n.d. Cultivo de la Yuca. Programa de Papa y Yuca, Instituto Colombiano Agropecuario, Tibaitata, Colombia. Mimeographed.

EWEL, JOHN J., and MADRIZ, ARNALDO
1968. *Zonas de vida de Venezuela*. Caracas: Ministerio de Agricultura y Cría.

F.A.O. STAFF
1957. Shifting Cultivation. *Unasylva*, Vol. XI, no. 1: 9–11.

FERNANDEZ Y., AGUSTIN, and MARTIN S., FELIPE
1953. Apuntes sobre la ictiología de Perijá. In *La región de Perijá y sus habitantes* [pp. 299–313]. Caracas: S.C.N.L.S.

FREEMAN, J. D.
1955. *Iban Agriculture: A Report on the Shifting Cultivation of Hill Rice by the Iban of Sarawak*. Colonial Research Studies. Vol. 18. London: H.M.S.O.

FREILE, ALFONZO J.
1968. Regiones climáticas de Venezuela. *Boletín de Geología*, Vol. X, no. 19: 3–156. Caracas: Min. de Minas e Hidrocarburos, Dirección de Geología.

GEERTZ, CLIFFORD
1966. *Agricultural Involution: The Process of Ecological Change in Indonesia*. Berkeley and Los Angeles: University of California Press.

GINÉS, HNO., AVELEDO, RAMÓN, et al.
1953. Lista y comentario de las aves colectadas en la Región. In *La región de Perijá y sus habitantes* [pp. 225–277]. Caracas: S.C.N.L.S.

Ginés, Hno., and Foldats, Ernesto
1953. Aspectos geobotánicos de la Región. In *La región de Perijá y sus habitantes* [pp. 327–340]. Caracas: S.C.N.L.S.

Ginés, Hno., Foldats, Ernesto, and Matos, Felipe
1953. Flórula de la cuenca del Río Negro, Perijá. In *La región de Perijá y sus habitantes* [pp. 345–551]. Caracas: S.C.N.L.S.

Ginés, Hno., and Wilbert, J.
1960. Una corta expedición a tierras motilonas. *Memoria de la Sociedad de Ciencias Naturales La Salle,* 20: 159–174.

Gourou, Pierre
1958. *The Tropical World.* Second edition. London: Longmans, Green and Company.

Grant, Ulysses J., Hatheway, William H., Timothy, David H., et al.
1963. *Races of Maize in Venezuela.* National Academy of Sciences–National Research Council Publ. 1136, Washington, D.C.

Gusinde, Martin
1956. The Yupa Indians in Western Venezuela. *Proceedings of the American Philosophical Society,* Vol. 100: 198–219.

Hanes, Leonard C.
1952. Phonemes of Motilone (Carib). *International Journal of American Linguistics,* 18: 146–149.

Harris, David R.
1971. The Ecology of Swidden Cultivation in the Upper Orinoco Rain Forest, Venezuela. *Geographical Review,* Vol. LXI, no. 4: 475–495.

Hea, James P. and Whitman, Adam B.
1960. Estratigráfia y petrología de los sedimentos pre-cretácios de la parte norte central de la Sierra de Perijá, Estado Zulia, Venezuela. In Tercer Congreso Geológico Venezolano, Vol. I. *Boletín de Geología.* [Caracas]. Publicatión Especial no. 3, pp. 351–376.

Hedberg, H. D., and Sass, L. C.
1937. Sinópsis de las formaciones geológicas de la parte occidental de la Cuenca de Maracaibo, Venezuela. *Boletín de Geología y Minas,* Tomo I, nos. 2, 3, and 4. Caracas: Min. de Fomento.

Higgs, J. W. Y., Kerkham, R. K., and Raeburn, J. R.
1950. *Report of a Survey of Problems in the Mechanization of Native Agriculture in Tropical African Colonies.* London: The Colonial Office.

Hildebrandt, Martha
1958. Sistema fonémico del Macoita. *Lenguas Indígenas de Venezuela,* no. 1. Caracas: Comisión Indigenista Venezolana.

Hitchcock, A. S.
1951. *Manual of the Grasses of the United States.* U.S.D.A. Misc. Publ. No. 200. Washington, D.C.: U.S. Government Printing Office.

Hitchcock, C. B.
1954. The Sierra de Perijá, Venezuela. *Geographical Review,* Vol. 44, no. 1: 1–28.

Holdridge, Leslie R.
1964. *Life Zone Ecology.* San José, Costa Rica: Tropical Science Center.

IZIKOWITZ, KARL G.
1951. *Lamet, Hill Peasants of French Indochina.* Etnologiska Studier 17. Goteborg: Etnografiska Museet.

JAHN, ALFREDO
1927. *Los aborígenes del occidente de Venezuela.* Caracas.

JENNINGS, D. L.
1970. Cassava in East Africa. *Tropical Root and Tuber Crops Tomorrow,* Vol. I: 64–65. Proceedings of the Second International Symposium on Tropical Root and Tuber Crops. Honolulu: College of Tropical Agriculture, University of Hawaii.

KAMPTO UTOMO
1957. *Masjarakat Transmigran Spontan Didaerah W. Sekampung (Lampung).* Djakarta: P. T. Penerbitan Universitas.

KOENS, A. J.
1948. *De Landbouw in de Indische Archipel.* Part II A: 165–200.

LABRADOR, JOSÉ RAMÓN
1967. *Estudios de biología y combate de gusano cogollero de maíz, Laphygma frugiperda (S and A).* Maracaibo: Publicaciones de la Facultad de Agronomía, Universidad del Zulia.

LAYRISSE, MIGUEL, LAYRISSE, ZULAY, and WILBERT, JOHANNES
1960. Blood Group Antigen Tests of the Yupa Indians of Venezuela. *American Anthropologist,* 62: 418–436.

LAYRISSE, M., and WILBERT, J.
1966. *Indian Societies of Venezuela: Their Blood Group Types.* Caracas: Editorial Sucre.

LEEDS, ANTHONY
1961. Yaruro Incipient Tropical Forest Horticulture: Possibilities and Limits. In Johannes Wilbert, ed., *The Evolution of Horticultural Systems in Native South America: Causes and Consequences.* Anthropológica Supplement, publ. no. 2. Caracas: S.C.N.L.S.

LIDDLE, RALPH A.
1946. *The Geology of Venezuela and Trinidad.* Ithaca, New York: The Paleontological Research Institution.

LODARES, BALTASAR DE
1930. *Los Franciscanos Capuchinos en Venezuela.* Caracas: Empresa Gutenberg.

MAGOON, M. L.
1970. Problems and Prospects in the Genetic Improvement of Cassava in India. *Tropical Root and Tuber Crops Tomorrow,* Vol. I: 58–61. Proceedings of the Second International Symposium on Tropical Root and Tuber Crops. Honolulu: College of Tropical Agriculture, University of Hawaii.

MASON, J. A.
1950. The Languages of South American Indians. In Julian Steward, ed., *The Handbook of South American Indians,* Vol. 6: 157–317. Washington, D.C.: Bureau of American Ethnology.

MENDEZ-AROCHA, A.
1957. Descripción de una colección de armas Chaké. *Antropológica*, 3: 49–57.
1958. Notas a la "Descripción de armas Chaké." *Antropológica*, 5: 39–52.

MÉTRAUX, ALFRED, and KIRCHHOFF, PAUL
1948. The Northeastern Extension of the Andean Culture. In Julian Steward, ed., *The Handbook of South American Indians*, Vol. 4: 349–368. Washington, D.C.: Bureau of American Ethnology.

MINISTERIO DE MINAS E HIDROCARBUROS
1956. *Léxico estratigráfico de Venezuela*. Publicación Especial No. 1.

NORMANHA, EDGARD S.
1970. General Aspects of Cassava Root Production in Brazil. *Tropical Root and Tuber Crops Tomorrow*, Vol I. 61–63. Proceedings of the Second International Symposium on Tropical Root and Tuber Crops. Honolulu: College of Tropical Agriculture, University of Hawaii.

NYE, P. H., and GREENLAND, D. J.
1965. *The Soil Under Shifting Cultivation*. Technical Communication No. 51. Commonwealth Bureau of Soils, Harpenden. Farmham Royal, England: Commonwealth Agricultural Bureau.

PARSONS, JAMES J.
1970. The "Africanization" of the New World Tropical Grasslands. In *Beitrage Zur Geographie de Tropen und Subtropen; Festschrift für Herbert Wilhelmy*. Tubingen: Tubinger Geographische Studien, no. 34.

PÉREZ A., ENRIQUE
1956. *Plantas útiles de Colombia*. Bogotá.

PETRICEKS, R. J.
n.d. *Shifting Cultivation in Venezuela*. Thesis, School of Forestry, Universidad de los Andes, Merida.

PHELPS, WILLIAM H.
1943. Las aves de Perijá. *Boletín de la Sociedad Venezolana de Ciencias Naturales*, 56: 265–338.

RADELLI, LUIGI
1962. Acerca de la geología de la Serranía de Perijá entre Codazzi y Villanueva (Magdalena–Guajira) Colombia. *Geología Colombiana*, no. 1: 23–41.

RAMOS, NELSON ESTRADA
1970. Colombian Studies for the Improvement of *Manihot Esculenta* Culture. *Tropical Root and Tuber Crops Tomorrow*, Vol. I: 83. Proceedings of the Second International Symposium on Tropical Root and Tuber Crops. Honolulu: College of Tropical Agriculture. University of Hawaii.

REICHEL-DOLMATOFF, GERARDO
1945. Los indios Motilones: etnografía y lingüística. *Revista del Instituto Etnológico Nacional* [Bogotá], Vol II, no. 1: 15–115.
1960. Contribuciones al conocimiento de las tribus de la Región de Perijá. *Revista Colombiana de Antropología* [Bogotá], Vol. 9: 159–198.

REICHEL-DOLMATOFF, GERARDO and ALEXANDER L. CLARK
1950. Parentesco, parentela y agresión entre los Iróka. *Journal de la Société des Americanistes* [Paris], 39: 15–57.

Renedo, Romualdo de
1969. De los Angeles del Tucucu a Sokorpa. *Venezuela Misionera,* Año XXXI [Junio], no. 362: 166–169.

República de Venezuela, Dirección General de Estadística y Censos Nacionales
1968. *Noveno censo general de población.* Caracas.

República de Venezuela, Dirección de Obras Hidráulicas
Anuario climatológico. Caracas: Ministerio de Obras Públicas.

Richards, P. W.
1957. *The Tropical Rain Forest.* Cambridge: The University Press.
1970. *The Life of the Jungle.* New York: McGraw–Hill.

Rivet, Paul and Armellada, C.
1950. Les Indiens Motilones. *Journal de la Société des Americanistes* [Paris], 39: 15–57.

Roberts, L. M., Grant, U. J., Hatheway, W. H., et al.
1957. *Races of Maize in Colombia.* National Academy of Sciences–National Research Council Publ. 510, Washington, D.C.

Rogers, David J.
1963. Studies of *Manihot esculenta* Crantz and related species. *Bulletin of the Torrey Botanical Club,* Vol. XL, No. 1: 43–54.

Ruddle, Kenneth
1970*a*. The Yukpa Autosubsistence System: A Study of Shifting Cultivation in Colombia and Venezuela. Ph.D. dissertation. Ann Arbor, Michigan: University Microfilms.
1970*b*. The Hunting Technology of the Maracá Indians. *Antropológica,* 25: 22–63.
1971. Notes on the Nomenclature and Distribution of the Yukpa-Yuko Tribe. *Antropológica,* 30: 18–28.
1973. The Human Use of Insects: Examples from the Yukpa. *Biotropica: Journal of the Association for Tropical Biology,* Vol. V, no. 2 [September]: pp. 94–101.

Santelos, Predencio de
1959–1960. Etnografía Yucpa: economía doméstica. *Venezuela Misionera.* [Caracas], nos. 247–252.

Sauer, C. O.
1952. *Agricultural Origins and Dispersals.* New York: American Geographical Society.

Schauensee, R. Meyer de
1964. *The Birds of Colombia.* Narberth, Pennsylvania: Livingston Publishing Company.

Schlippe, Pierre de
1955. *Shifting Cultivation in Africa: The Zande System of Agriculture.* London: Routledge and Kegan Paul.

Schnee, L.
1960. Plantas comunes de Venezuela. *Revista de la Facultad de Agronomía,* [Maracay], No. 3 [October].

Schön, Miguel, and Jam, Pedro

1953a. Economía. In *La Regíon de Perijá y sus habitantes,* 5: 56–69. Caracas: S.C.N.L.S.

1953b. Cultura social. In *La Regíon de Perijá y sus habitantes,* 5: 71–94. Caracas: S.C.N.L.S.

Sievers, Wilhelm

1888. Die Sierra Nevada de Santa Marta und die Sierra de Perijá. *Zeitschrift der Gesellschaft für Erdkunde,* [Berlin], Vol. 23: 1–158.

Simón, Pedro de

1882–1892. *Noticias historiales de la conquista de tierra firme en las Indias Occidentales.* Bogotá.

Simon, F. A. A.

1881. On the Sierra Nevada and Its Watershed. *Proceedings of the Royal Geographical Society,* Vol. 3: 705–723.

Sociedad de Ciencias Naturales La Salle

1953. *La región de Perijá y sus habitantes.* Publicaciones de la Universidad del Zulia. Caracas. Editorial Sucre.

Spencer, Joseph E.

1966. *Shifting Cultivation in Southeastern Asia.* University of California Publications in Geography. Vol. 19. Berkeley and Los Angeles: University of California Press

Spencer, J. E., and Stewart, N. R.

1973. The Nature of Agricultural Systems. In *Annals,* The Association of American Geographers, Vol. 63, No. 4, pp. 529–544.

Tax, Sol

1953. *Penny Capitalism: A Guatemalan Indian Economy.* Smithsonian Institution–Institute of Social Anthropology Publication No. 16, Washington, D.C.

Uscategui Mendoza, Nestor

1961. Distribución actual de las plantas narcóticas y estimulantes usadas por las tribus indígenas de Colombia. *Revista de la Academia de Ciencias Exactas, Físicas y Naturales,* [Bogotá], Vol. II, no. 43: 215–228.

Vila, Pablo

1965. *La geografía de Venezuela.* Caracas: Ministerio de Educación. Two volumes.

Villamañan, Adolfo, and Santelos, Prudencio de

1956–1957. Los Japreria: quienes son y donde viven. *Venezuela Misionera* [Caracas], Año 18: Nos. 209–215, Año 19: Nos. 217–219.

Watters, R. F.

1960a. Some Forms of Shifting Cultivation in the South West Pacific. *The Journal of Tropical Geography* [University of Malaya], Vol. 14: 35–50.

1960b. The Nature of Shifting Cultivation. *Pacific Viewpoint,* Vol. I, no. 1:59–99.

1965. *Shifting Cultivation in Venezuela.* Manuscript for F.A.O. Report, Forestry and Forest Products Division.

1971. *Shifting Cultivation in South America.* Forestry Development Paper. Paper No. 17. Rome: F.A.O.

WILBERT, JOHANNES

1960. Zur Kenntnis der Parirí. *Archiv für Volkerkunde* [Vienna], Vol. 15: 80–153.

1961. Identificación étno-lingüística de las tribus indígenas del occidente de Venezuela. *Memoria de la Sociedad de Ciencias Naturales La Salle,* 58: 5–27.

1974. *Yupa Folktales.* Latin American Studies Series, Vol. 24. Los Angeles: Latin American Center, University of California.

WOKITTEL, ROBERTO

1957. Bosquejo geográfico y geológico de la Sierra Nevada de Santa Marta y la Serranía de Perijá. *Boletín Geológico* [Bogotá], Vol. 5, no. 3: 29–50.

YEPES, A. F.

1945. Anotaciones sobre los indios Rionegrinos de Perijá, Venezuela. *Acta Americana,* Vol. 3, nos. and 2: 64–70.

INDEX

PLATES

Plantains and bananas planted on a terrace being undercut by floodwaters.

A scaffold (*tapóymotos*) constructed to fell a large tree above the buttress.

a, A *tíšrane* in which only the *nikókta* has been consumed.

b, Detail of an unevenly burned field where some of the larger trunks are well-burned whereas much of the smaller material remains uncharred.

Čáčo's bean field shortly after burning in 1969, the first year of use. The rockiness of the terrain, incipient gullying, and invasion by stoloniferous grasses (top right) are evident. The field burned badly resulting in a considerable reduction of the plantable area.

a, The impact of escaped field burns at Socorpa. Secondary forest, surrounded by severely burned and dead trees, is confined to the valleys. Ridges support impoverished fern and grass associations.

b, Repeated fire drives to hunt mammals and insects have caused severely pauperized grassland soils and limit tree growth at Manastará (Negro Valley).

Crib (*karáčašk*) for storing shelled maize.

Crib (*tipértartos*) for storing unshelled maize.

a, Carefully weeded field of young yautia (*Xanthosoma sagittifolium*).

b, Common millet (*Panicum miliaceum*) sown in clumps in a dooryard garden for the production of arrow shafts.

Yam plant trained to climb a stump in a dooryard garden.

Trellis (*tapótritos*) constructed alongside a dwelling to support beans.

Section of a field in the second year of cultivation. Visible crops are canna, yautia, papaya and tobacco.

a, Section of a field in second year of cultivation devoted to manioc (*Manihot dulcis*). Cuttings recently replanted after harvesting of mature crop appear in the foreground.

b, Section of a dooryard garden showing gourds (*Lagenaria* sp.) drying on the vine, tobacco, and castor bean.

Part of a *títrano* in which several hectares of high forest fallow were destroyed. The repeatedly burned ridge in the background supports fern-dominated associations.

a, Dooryard garden at Socorpa with tobacco and squash sown close to the dwelling. The stump and trunk provide a convenient source of firewood.

b, A more elaborate dooryard garden at Socorpa in which field crops have been planted. Amaranth (*Amaranthus cruentus*) dominates the foreground.

A Maracá extended family compound surrounded by *rastrojo* about five years old. In the background is an abrupt, rocky escarpment (*tókru*) under tall secondary forest.

Ewa's *finca* in El Valle de los Motilones. In the foreground coffee, plantains, and bananas are interplanted. Pineapple dominates the middle ground above the house compound and the calf corral. The upper area is under pasture and first year *rastrojo*.

a, Permanent occupance of a temporary hunting shelter (headwaters of the Tukuku).

b, Palisaded family dwelling (Socorpa).

c, Extended family compound at El Tukuku. Note *chicha* trough (*kanówa*) in foreground.

d, Ewa's house and garden in the mission village at El Tukuku.

a, Shelling seed corn prior to sowing.

b, Using the *coa* to interplant manioc between rows of maize in a new field.

c, Clearing one year's growth of *rastrojo* to plant yautia and plantains.

d, Making rope from Agave fibers.